# THE MYTH OF MONOGAMY

# THE
# MYTH OF
# MONOGAMY

## Fidelity and Infidelity in Animals and People

DAVID P. BARASH, Ph.D.
JUDITH EVE LIPTON, M.D.

W. H. Freeman and Company
New York

*Text Designer: Victoria Tomaselli*

Library of Congress Cataloging-in-Publication Data.

Barash, David P.
  The myth of monogamy : fidelity and infidelity in animals and people /
  David P. Barash and Judith Eve Lipton.
      p.    cm.
  Includes bibliographical references and index.
  ISBN 0-7167-4004-4
  1. Adultery.  2. Sex customs.  3. Sexual behavior in animals.  I. Title.

HQ806.B367 2001
                                                        2001023209

Printed in the United States of America

First printing 2001

To Ilona and Nellie,
for helping to make it all worthwhile

# Contents

# Acknowledgments

We have been working together for twenty-five years, and we are grateful to each other for deep friendship and wonderful evenings. We thank our editor John Michel for seeing merit in this project and for helping to improve it as it developed. Project editor Jane O'Neill was also crucial in making this book become a reality. We offer particular gratitude to Nellie Barash, our fifteen-year-old daughter, who assisted in correcting page proofs and slang. Thank you, Nellie, for being our culture maven. We want most of all to thank the many researchers who populate the pages of this text and the endnotes for the diligence and insight that has helped expose the myth of monogamy. And we offer our encouragement to everyone—critters, scientists, and lay people—who struggle with love and betrayal.

"The world is not to be narrowed till it will go into the understanding . . . but the understanding is to be expanded till it can take in the world."

— FRANCIS BACON (1561–1626)

# Monogamy for Beginners

A nthropologist Margaret Mead once suggested that monogamy is the hardest of all human marital arrangements. It is also one of the rarest. Even long-married, faithful couples are new at monogamy, whether they realize it or not. In attempting to maintain a social and sexual bond consisting exclusively of one man and one woman, aspiring monogamists are going against some of the deepest-seated evolutionary inclinations with which biology has endowed most creatures, *Homo sapiens* included. As we shall see, there is powerful evidence that human beings are not "naturally" monogamous, as well as proof that many animals, once thought to be monogamous, are not. To be sure, human beings *can* be monogamous (and it is another question altogether whether we *should be*), but make no mistake: It is unusual—and difficult.

As G. K. Chesterton once observed about Christianity, the ideal of monogamy hasn't so much been tried and found wanting; rather, it has been found difficult and often left untried. Or at least, not tried for very long.

The fault—if fault there be—lies less in society than in ourselves and our biology. Thus, monogamy has been prescribed for most of us by American society and by Western tradition generally; the rules as officially stated are pretty clear. We are supposed to conduct our romantic and sexual lives one-on-one, within the designated matrimonial playing field. But as in soccer or football, sometimes people go out of bounds. And not uncommonly, there is a penalty assessed if the violation is detected by a referee. For many people, monogamy and morality are synonymous. Marriage is the ultimate sanction and departures from marital monogamy are the ultimate interpersonal sin. In the acerbic words of George Bernard Shaw, "Morality consists of suspecting other people of not being legally married."

Ironically, however, monogamy itself isn't nearly as uncomfortable as are the consequences of straying from it, even, in many cases, if no one finds out. Religious qualms aside, the anguish of personal transgression can be intense (at least in much of the Western world), and those especially imbued with the myth of monogamy often find themselves beset with guilt, doomed like characters from a Puritan cautionary tale to scrub eternally and without avail at their adultery-stained souls, often believing that their transgression is not only unforgivable, but unnatural. For many others—probably the majority—there is regret and guilt aplenty in simply feeling sexual desire for someone other than one's spouse, even if such feelings are never acted upon. When Jesus famously observed that to lust after another is to commit adultery in one's heart, he echoed and reinforced the myth of monogamy—the often-unspoken assertion that even desire-at-a-distance is not only wrong, but a uniquely human sin.

Whether such inclinations are wrong is a difficult, and perhaps unanswerable, question. But as we shall see, thanks to recent developments in evolutionary biology combined with the latest in technology, there is simply no question whether sexual desire for multiple partners is "natural." It is. Similarly, there is simply no question of monogamy being "natural." It isn't.

Social conservatives like to point out what they see as a growing threat to "family values." But they don't have the slightest idea how great that threat really is or where it comes from. The monogamous family is very definitely under siege, and not by government, not by a declining moral fiber, and certainly not by some vast homosexual agenda . . . but by the dictates of biology itself. Infants have their infancy. And adults? Adultery.

If, as Ezra Pound once (somewhat self-servingly) observed, artists are the "antennae of the race," these antennae have long been twitching about extramarital affairs. If literature is any reflection of human concerns, then infidelity has been one of humankind's most compelling, long before biologists had anything to say about it. The first great work of Western literature, Homer's *Iliad,* recounts the consequences of adultery: Helen's face launched a thousand ships and changed the course of history only after it first launched an affair between Helen, a married woman and Greek queen, and Paris, son of King Priam of Troy. Helen proceeded to leave her husband Menelaus, thereby precipitating the Trojan War. And in the *Odyssey,* we learn of Ulysses' return from that war, whereupon he slays a virtual army of suitors, each of whom was trying to seduce his faithful wife, Penelope. (By contrast, incidentally, Ulysses himself had dallied with Circe the sorceress, but he was not considered an adulterer as a result. The double standard is ancient and by definition unfair; yet it, too, is rooted in biology.)

It seems that every great literary tradition, at least in the Western world, finds it especially fascinating to explore monogamy's failures: Tolstoy's *Anna Karenina,* Flaubert's *Madame Bovary,* Lawrence's *Lady Chatterley's Lover,* Hawthorne's *The Scarlet Letter,* Henry James's *The Golden Bowl.* More recently, John Updike's marriage novels—not to mention scores of soap operas and movies—describe a succession of suburban, middle-class affairs. The present book, by contrast, is not fiction. And it is not concerned with affairs as such, but rather with the biological underpinnings of affairs, in human beings and other animals as well. More precisely, it is about what the latest research has been revealing about the surprisingly weak biological underpinnings of monogamy.

Our approach will be biological, because whatever else human beings may be, we are biological creatures through and through. We eat, we sleep, we feel emotions, we engage in sex, and although we are unique in some regards, so is every other living thing! Rhinos and cobras are uniquely rhinos and cobras in their evolutionary history, their physiology, their anatomy, their behavior, just as human beings are uniquely human. But are we—can we be—*more unique* than other creatures? Moreover, it should quickly become apparent that despite the oxymoronic "shared uniqueness" of all living things, there is also a genuine commonality of pattern, especially—for our purposes—a shared susceptibility to certain basic behavioral tendencies. It is taken for granted that we can learn about human digestion, respiration, or metabolism by studying these processes in other animals, making due allowance, of course, for certain unavoidable differences among different species. The same applies for much, although assuredly not all, behavior.

In this book, we'll be concerned with a range of living things, in part because each is worth understanding in its own right and also because of the light they can shed on ourselves. Don't misunderstand: There will be no claim that because hairy-nosed wombats show a particular sexual pattern, people do, too. Arguments of this sort are absurdly naive, if only because there is such remarkable variety in the animal world. Among so-called lekking species of birds, for example, males gather at a ceremonial plot of ground, with each male defending a small territory; many different females then mate preferentially with one of these males, typically the one occupying the most central "lek" and whose displays are especially intense. (No pair-bond here.) Then there are the pygmy chimps, also known as bonobos, which engage in what seems to be a nonstop sexual free-for-all. Once again, nothing close to monogamy is found . . . and these are our closest animal relatives.

On the other hand, there are cases of lifelong social and sexual partnership that might give pause even to the most committed advocates of intimate, intense, altogether faithful male–female bonding: Not many living things partake, for example, of the extreme monogamy shown by the parasitic flatworm *Diplozoon paradoxum,* a fish parasite whose partners meet as virgin adolescent larvae, whereupon they literally fuse at their midsections and subsequently become sexually mature; they then remain "together" (in every sense of the word) till death parts them—in some cases, years later.

The above examples ranged from birds to mammals to invertebrates. And yet it isn't at all clear which is most "relevant" to human beings. If by relevant we mean which one provides a model or—worse yet—a set of rules or some sort of evolutionary premonition as to our "deeper selves," the answer must be: none. But at the same time, each is relevant in its own way. Not only does every animal species cast its unique light on life's possibilities, but each case also helps illuminate a facet of ourselves.

For most laypersons, there is an understandable bias toward mammals, especially primates. But although the lives of chimpanzees, gorillas, gibbons, and orangutans are fascinating and colorful (especially the highly sexed bonobos, about whom more later), the truth is that when it comes to similarities between their lives and those of human beings, these great apes just aren't that great. Birds—at least, certain species—are far more informative.

This is because we're not looking for direct historical antecedents, but rather for similarities based on similar circumstances. Among nearly all mammals, including most primates, monogamy just isn't in the picture. Nor is male care of the young. By contrast, although birds aren't nearly as monogamous as once thought, they are at least inclined in that direction. (The same can be said about human beings.) Not only that, but social monogamy—as opposed to genetic monogamy—is strongly correlated with parental involvement on the part of fathers as well as mothers, a situation that is common in birds and quite unusual among mammals, except for that most birdlike of primates, *Homo sapiens.*

In this book, we won't be especially focused on mammals (except for ourselves). When it comes to dispelling the myth of monogamy, most of the really useful discoveries in recent years have come from research by ornithologists, who, interestingly enough, have directed much of their attention to those species that are "polygynous" (where the typical mating arrangement is one male and many females) or "polyandrous" (one female and many males). Only recently have they turned their attention to monogamy, only to discover that it is more myth than reality.

We'll also spend some time with invertebrates, because they include so many different species, each of which is, in a sense, a distinct zoological experiment, whose results we are only now beginning to decipher.

Certain insects have had an important historical role in helping us appreciate the rarity of monogamy. Thus, some time ago, environmentalists had great hope for a novel technique that promised to eradicate insect pests. The idea was to release large numbers of sterilized males, which would mate with females, who would therefore fail to reproduce. Eventually, no more pests . . . and no more pesticides, to boot. But the success of this procedure never extended beyond one species, the screw-worm fly.

This is what happened. During the 1930s, E. F. Knipling, a forward-looking entomologist with the U.S. Department of Agriculture, may have sensed that "natural" (that is, noninsecticidal) means of controlling unwanted insects would be superior to the widespread use of poisons. In any event, he began exploring a promising technique: Introduce sterilized male screw-worms into nature, whereupon they would mate with wild female screw-worms, whose offspring would fail to materialize. It worked, becoming for a time one of the great success stories of post–Rachel Carson environmentalism. By the 1960s, male screw-worms were being exposed to radioactive cobalt by the vatful, after which insect eunuchs were airdropped over a vast region along the Mexican–U.S. border. This technique succeeded in eliminating the screw-worm scourge. However, such an outcome has never been replicated. As it turns out, Knipling's choice of a target species was fortunate (or scientifically inspired): Female screw-worms—despite their name—are strictly monogamous. By contrast, we now know that for nearly all insects, one screw is not enough: Females commonly mate with more than one male, so even when they are inundated with a blizzard of sterile males, it only takes a small number of intact ones for reproduction to go merrily along. And so the "sterile-male technique," for all its environmental, nonpesticide appeal, has gone nowhere.

At the same time, the door was opened to a startling insight—namely, that multiple mating is common in nature. And here is the key point: Multiple mating doesn't refer only to the well-known tendency of males to seek numerous sexual partners, but to females, too. Probably the first modern biologist to call attention to this phenomenon, and to recognize its significance, was British behavioral ecologist Geoffrey A. Parker. In 1970, in what can truly be called a seminal paper, Parker wrote of "Sperm Competition and Its Evolutionary Consequences in the Insects." In one stroke, a new idea was born (or, at least, recognized). It is really a simple concept, a direct result, in fact, of multiple mating: Sperm from more than one male will often compete to fertilize a female's eggs. Sperm competition is in no way limited to insects; examples have been found in virtually every animal group . . . including human beings.

Sperm competition is essentially another way of saying nonmonogamy. If a female mates with only one male, then, by definition, no sperm competition occurs. (Except, of course, for the scramble among individual pollywogs within an ejaculate. Although this may be intense, it is nonetheless different from competition among sperm from different males.) Another way of saying this: If females mate with more than one male, sperm competition will ensue. Of course, this depends on the females in question being non-monogamous, something we can prove by showing that their offspring were sired by more than one male.

Sperm competition was actually first documented by none other than Charles Darwin, although he did not identify it as such. Indeed, Darwin seems to have carefully refrained from pursuing the matter, perhaps because the question of females mating with more than one male was more than Darwin's social climate could bear. Thus, in *The Descent of Man and Selection in Relation to Sex* (1871), Darwin described a female domestic goose who produced a mixed brood consisting of some goslings fathered by a domestic goose who was her social partner as well as others evidently fathered by a Chinese goose . . . this second male being not only not her mate, but also not even of the same species!

Darwin's refusal to pursue the question of extra-pair copulations—those occurring outside the ostensibly monogamous pair-bond—may have been more than simply a quaint Victorian fastidiousness. Even today, in our supposedly liberated sexual climate, many people get a bit queasy over the image of sperm from more than one man competing within the vagina and uterus of a single woman. ("Single," that is, as opposed to plural; such a woman may well be married or otherwise paired with an identified male: That is the point.)

Here is an account of sperm competition, co-authored by the doughty Geoffrey Parker and intended to present the basic points of a simple physical model, known as "constant random displacement with instant sperm mixing."

Imagine a tank of sperm, representing the fertilization set, which has an input pipe and an outlet pipe. During copulation, sperm flow at a constant rate into the tank through the inlet and out (by displacement) through the outlet. First imagine that the sperm entering the tank do not mix with the sperm already present, which are pushed towards and out from the outlet. The new sperm displace only the old sperm, so that the proportion of sperm from the last male . . . rises linearly at a rate equal to the input rate. . . . But, now suppose that there is swift random mixing of the oncoming sperm with the previous sperm in the tank. At first the sperm displaced from the outlet will be only the old sperm. As the last male's sperm build up in the tank, some of the displaced sperm will be his own ("self-displacement"). By the time most of the sperm in the tank is new, most of the outflow will represent self-displacement.

Parker's physical model (accompanied with equations, predictions, and supportive data) is entirely sound and logical. At the same time, the very idea of a tankful of sperm is not one likely to make the heart sing! (Why not? I'm not at all sure, but it wouldn't be surprising if most people's disinclination to think deeply and cheerfully about semen or sperm is related to the general disinclination of biologists to think about non-monogamy among animals and, in turn, to the discomfort most of us feel when considering non-monogamy among human beings as well. Not to mention a likely female disinclination to be designated a "tank"!)

Geoffrey Parker's initial studies of sperm competition employed the "irradiated-male" technique, much like Knipling's more applied research several decades earlier. In Parker's case, the idea was that after subjecting males to radiation, their sperm was damaged, not enough to prevent them from fertilizing eggs but sufficient to interfere with the normal development of any resulting embryo. So, by mating females to irradiated and nonirradiated males, then counting how many eggs were fertilized but didn't develop in each case, it was possible to assign paternity and thus calculate the success of the different males.

Other techniques quickly followed, notably direct genetic evidence for multiple paternity using "allozymes." This depends on the existence of distinct genetic differences among individuals. In some cases, these differences are well known and readily seen, as with traits such as eye color or hair color in people, the presence or absence of attached earlobes, or the ability or inability to roll one's tongue. For most allozyme-based studies, however, the genetic differences in question are more subtle, analogous to blood types. Knowing, for example, whether a child's blood type is A, B, AB, or O, it is possible to determine whether a given adult could have been the father. (For example, if a child is type O, a man accused in a paternity suit could not be the father if he is type AB.) But it is one thing to prove or disprove the *possibility* of parentage—to say that someone *could* or, alternatively, *could not* be the father—and quite another to say that he definitely *is*.

Such certainty is now available. It required the next and most significant breakthrough to date on the way toward disproving the myth of monogamy: the discovery of "DNA fingerprinting," not only for human beings but also animals. Just as each person has a unique fingerprint, each of us has a unique pattern of DNA, so-called minisatellite regions that are "hypervariable," offering a range of possibility that encompasses more than a hundred million different identifying traits, far more than, for example, blood types A, B, AB, or O. As a result, just as each citizen of the United States can be uniquely identified by a personalized Social Security number (so long as we allow enough digits), DNA fingerprinting provides enough genetic specification to guarantee that only one individual will possess a particular pattern.

Given tissue samples from offspring and adults, we can now specify, with certainty, whether a particular individual is or is not the parent, just as it is possible to specify, with certainty, the donor of any sample of blood, hair, or semen. After subjecting the tissue to appropriate treatments, research technicians end up with a DNA profile that looks remarkably like a supermarket bar code, and with about this level of unique identification. Armed with this technique, field biologists—studying the behavior of free-living animals in nature—have at long last been able to pinpoint parenthood. As a result, the field of "biomolecular behavioral ecology" has really taken off, and with it, our understanding of a difference that may sound trivial but is actually profound: between "social monogamy" and "sexual monogamy."

Two individuals are socially monogamous if they live together, nest together, forage together, and copulate together. Seeing all this togetherness, biologists not surprisingly used to assume that the animals they studied were also mixing their genes together, that the offspring they reared (usually together) were theirs and theirs alone. But thanks to DNA fingerprinting, we have been learning that it ain't necessarily so. Animals—not unlike people—sometimes fool around, and much more often than had been thought. When it comes to actual reproduction, even bird species long considered the epitome of social monogamy, and thus previously known for their fidelity, are now being revealed as sexual adventurers. Or at least, as sexually non-monogamous.

Incidentally, it is not easy to obtain the so-called minisatellite DNA profiles needed to assign accurate parentage to animals—or to human beings, for that matter. The actual laboratory techniques are elaborate and detailed. Here is a sample, taken from the "methodology" section of a recent scientific paper describing this latest wedding of genetic insight to animal sexual behavior. We present it here not to provide a cookbook recipe for do-it-yourself DNA analysts, but as a kind of penance, so that when in this or subsequent chapters you encounter an off-hand mention of "DNA fingerprinting," you will pause—if only briefly—and give credit to the sophisticated labor that made such information possible:

> We added 30 μl of 10% SDS and 30 μl of proteinase K and incubated the [blood] sample at 55 degrees C for 3 h. A further 10 μl of proteinase K were then added and the sample returned to 55 degrees C overnight. An extra Tris-buffered Phenol wash was also performed to remove additional proteins present in the tissue. To 20 μl of genomic DNA, we added 4 μl 10 × Buffer (react 2), 2 μl 2 mg/ml BSA, 1 μl 160 nM Spoermidine, 1 μl of the restriction enzyme

*Hae*III and 11 μl milli-Q water. This mixture was incubated overnight at 37 degrees C. Another 1 μl of *Hae*III was added the following day and the sample was incubated at 37 degrees C for a further 1 h. Digested samples were then stored at –20 degrees C. About 5 μg of digested DNA were loaded in each lane of the gel. DNA fragments were resolved on a 0.8% agarose gel (19 × 27 cm) in 1 × TBE running buffer at 55 C for 72 h. We then denatured the DNA by washing each gel for 15 min in 0.25 M HCl and then for 45 min in 0.5 M NaOH, 1.5 M NaCl. Gels were then neutralized by two 15-min washes in 1.5 M NaCl, 0.5 M Tris-HCl pH 7.2, 1mM EDTA. Southern blot techniques were used to transfer DNA from agarose gels to nulon membranes in 6 × SSC. Membranes were then dried for 10 min at 37 degrees C before being baked at 80 degrees C for 2 h.

Baked membranes were soaked in prehybridization mix (75 ml 0.5 disodium hydrogen orthophosphate pH 7.2. 75 ml milli-Q water, 300 μl 0.5 M EDTA pH 8.0, 10.5 g SDS for 2 h at 65 degrees C. First a Jeffrys 33.15 probe was labelled with a-$^{32}$ PdCtp by random priming with Amersham radprime kit. Unincorporated label was removed using a G50 sephadex column. Hybridization of Jeffreys 33.15 to membranes was at 65 degrees C for a minimum of 18 h. Membranes were then washed twice with 5 × SSC, 0.1% SDS at 65 degrees C. DNA fragments hybridized to the 33.15 probe were exposed on X-ray film at either –80 degrees C with one intensifying screen or at room temperature for 1–6 days. After adequate exposure, membranes were stripped and reprobed with CA probe, which was similarly labelled to a-$^{32}$ PdCTP.

Had enough?

*M*onogamy generally implies mating exclusivity. In this book, we shall use the term to mean a social system in which the reproductive arrangement appears to involve one male and one female. But the burden of our argument is that when it comes to monogamy as mating exclusivity, what we see is not necessarily what we get. Herein lies the myth.

When asked, men consistently claim to have had more sexual partners than women. As we shall see, it is consistent with evolutionary theory that when it comes to sex, males are comparatively indiscriminate whereas females are likely to be more careful and cautious. But this is only possible if a small number of women make themselves sexually available to a large number of men, because, assuming that every heterosexual encounter involves one man and one woman, the numbers must balance out. There is also strong evidence that men tend to exaggerate their reported number of sexual encounters, while women tend to understate theirs. This discrepancy could result from genuine memory lapses on the part of women and/or unconscious deception (of self and others) by members of both sexes. In addition, social pressures prescribe that having multiple sexual partners (over time, mind you, not necessarily simultaneously) indicates a "real man," whereas being a "real"—that is, virtuous—woman has long been equated with monogamous fidelity. In any event, it is interesting that among many animals, too, females are especially secretive about their extra-pair copulations, whereas males are comparatively brazen . . . even if they are not inclined to verbal exaggeration.

People (and not just scientists!) have long known that the human species is prone to more than a bit of hypocrisy, saying one thing with regard to fidelity and then—at least on occasion—doing another. But when it comes to the scientific study of animal mating systems, biologists had traditionally assumed that when a species "is" socially monogamous, then it really *is* monogamous; that is, sexually exclusive. No more.

In the movie *Heartburn*, a barely fictionalized account by Nora Ephron of her marriage to the philandering Carl Bernstein, the heroine (played by Meryl Streep) tearfully tells her father about her husband's infidelities, only to be advised, "You want monogamy? Marry a swan." But now, it appears that not even swans are reliably monogamous.

Reports of extra-pair copulations in animals previously thought to be monogamous have come hot and heavy during the last decade or so. Increasingly, biology journals have featured articles with titles such as "Behavioral, Demographic and Environmental Correlates of Extra-Pair Fertilizations in Eastern Bluebirds," "Multiple Paternity in a Wild Population of Mallards," "Extrapair Copulations in the Mating System of the White Ibis," "DNA Fingerprinting Reveals Multiple Paternity in Families of Great and Blue Tits," "Extrapair Paternity in the Shag, as Determined by DNA Fingerprinting," "Genetic Evidence for Multiple Parentage in Eastern Kingbirds," "Extra-Pair Paternity in the Black-Capped Chickadee," "Density-Dependent Extra-Pair Copulations in the Swallow," "Patterns of Extra-Pair Fertilizations in Bobolinks," and "Extra-Pair Paternity in Monogamous Tree Swallows." We have even had this oxymoronic report: "Promiscuity in Monogamous Colonial Birds."

The situation has reached the point where *failure* to find extra-pair copulations in ostensibly monogamous species—that is, cases in which monogamous species *really* turn out to be monogamous—is itself reportable, leading to the occasional appearance of such reassuring accounts as "DNA Fingerprinting Reveals a Low Incidence of Extra-Pair Fertilizations in the Lesser Kestrel," or "Genetic Evidence for Monogamy in the Cooperatively Breeding Red-Cockaded Woodpecker."

Until recently, for a scientific journal to publish a report demonstrating that a "monogamous" species is in fact monogamous would be as silly as for it to publish an account revealing that a particular species of mammal lactates and nurses its young. "Big deal," its readers would say. But now, with the tidal wave of evidence for genetic non-monogamy, any evidence for true monogamy is a big deal indeed, even among those bird species such as eagles and geese that were long seen as paragons of pair-bonding.

The plot thickens: When migrating birds were live-trapped and the cloacas of the females rinsed out and examined, at least 25 percent of them were revealed to be already carrying sperm. And this *before* having reached the breeding areas to which they were headed! Evidently, when females—even young ones, in their first reproductive year—arrive at their breeding areas and set up housekeeping with a territorial male, more than a few have already lost their virginity. The likelihood is that such sexual experiences are nonfunctional, or at least nonreproductive, although this remains to be proven, since live sperm can be stored for several days within the genital tract of most birds.

In any event, it is difficult to overstate the conceptual revolution that has followed the discovery that copulations—and, in many cases, fertilizations—often take place outside the social unions that researchers typically identify. After all, reproductive success is the fundamental currency of evolutionary success, and behavioral ecologists and sociobiologists studying red-winged blackbirds, for example, have long been in the habit of evaluating the reproductive success of their male subjects by counting harem size or, better yet, the number of young birds produced by all of a male's "wives." But now comes word that in this polygynous species, too, females don't restrict their mating to the harem-keeper. It turns out that there is no necessary correlation between a male red-winged blackbird's apparent reproductive success (the number of offspring reared on his territory) and his actual reproductive success (the number of offspring he fathered). Similarly, there is no guaranteed correlation between his harem size and his actual reproductive success: A male red-winged blackbird (like a male Turkish sultan) can "have" many wives, which in turn can have many offspring—but those children might not be his.

The pattern is painfully clear: In the animal world generally, and the avian world in particular, there is a whole lot more screwing around than we had thought. (As to the human world, most people have long known that there is a whole lot more of the same than is publicly—or even privately—acknowledged.)

When it comes to mammals, monogamy has long been known as a rarity. Out of 4,000 mammal species, no more than a few dozen form reliable pair-bonds, although in many cases it is hard to characterize them with certainty because the social and sexual lives of mammals tend to be more furtive than those of birds. Monogamous mammals are most likely to be bats (a few species only), certain canids (especially foxes), a few primates (notably the tiny New World monkeys known as marmosets and tamarins), a handful of mice and rats, several odd-sounding South American rodents (agoutis, pacas, acouchis, maras), the giant otter of South America, the northern beaver, a handful of species of seals, and a couple of small African antelopes (duikers, dik-diks, and klipspringers). A pitiful list.

Even females of seemingly solitary species such as orangutans, gibbons, and black bears have been found to copulate with more than one male; hence, observations of social organization alone clearly can be misleading. Until recently, lacking the appropriate genetic techniques, we had little choice but to define monogamy by the social relationships involved; only with the explosion in DNA fingerprinting technology have we started to examine the genetic connections, those most important to evolution. Thus, according to the highly respected book by David Lack, *Ecological Adaptations for Breeding in Birds,* fully 92 percent of bird species are monogamous. Socially, this figure is still accurate; sexually, it is way off. The highest known frequency of extra-pair copulations are found among the fairy-wrens, lovely tropical creatures technically known as *Malurus spendens* and *Malurus cyaneus.* More than 65 percent of all fairy-wren chicks are fathered by males outside the supposed breeding group. Here is another eye-opener. Warblers and tree swallows are purportedly monogamous, yet when genetic analyses were conducted on six different offspring in each of these species, they were found to have been fathered by five different males!

Although such cases are admittedly extreme, we now know that it is not uncommon for 10 to 40 percent of the offspring in "monogamous" birds to be fathered by an "extra-pair" male; that is, one who isn't the identified social mate of the female in question. (It is much less common for offspring to be "mothered" by an extra-pair female; that is, for an outsider female to slip one of her eggs into the nest of a mated pair. More on this later.)

Given how much we have been learning about non-monogamy and extra-pair matings among animals, and considering the newfound availability of such testing, it is remarkable how rarely genetic paternity tests have been run on human beings. On the other hand, considering the inflammatory potential of the results, as well as, perhaps, a hesitancy to open such a Pandora's box, maybe *Homo sapiens'* reluctance to test themselves for paternity is sapient indeed. Even prior to DNA fingerprinting, blood-group studies in England found the purported father to be the genetic father about 94 percent of the time; this means that for six out of a hundred people, someone other than the man who raised them is the genetic father. In response to surveys, between 25 percent and 50 percent of U.S. men report having had at least one episode of extramarital sex. The numbers for women are a bit lower—around 30 percent—but still in the same ballpark. Many people already know quite a lot—probably more than they would choose to know—about the painful and disruptive effects of extramarital *sex*. It wouldn't be surprising if a majority would rather not know anything more about its possible genetic consequences, extramarital *fatherhood*. Maybe ignorance is bliss. (If you feel this way, better stop reading here!)

Until quite recently, multiple mating was hidden from biologists. It wasn't so much invisible as unacknowledged, a perfect example of the phenomenon that even in such a seemingly hard-headed pursuit as science, believing is seeing. More to the point, not believing is not seeing. Sexual infidelities among ostensibly monogamous species, when noticed at all by biologists, were generally written off as aberrant, not worth describing, and certainly not suitable for analysis or serious theory. Distasteful as it may have been, Geoffrey Parker's work changed that, along with this important recognition by evolutionary theoretician Robert Trivers. A "mixed strategy" should be favored, as least among males: Maintain a pair-bond with a female, whom you might well assist in rearing offspring, but be ready and available for additional copulations if the opportunity arises. The next step was to ask: What about the female? Is she merely a passive recipient of male attentions, an empty tank to be filled with the sperm of various competing paramours? Or does she choose among the eager male prospects? Might she even actively solicit extra-pair copulations, generating sperm competition among different males? And is it beyond the evolutionary ingenuity of females for them to play hard-to-get *within* their own genital tracts?

Early work, both empirical research and theorizing, took a decidedly male-centered perspective on multiple mating, emphasizing how males maximize their paternity by being sexually available to more than one female whenever possible, by competing with each other directly (via bluffing, displaying, and fighting) and indirectly (via guarding their mates), and by using an array of anatomical, physiological, and behavioral techniques—such as frequent copulations—to give them an advantage over other males. In his research, David, too, was guilty of this short-sightedness.

More recently, however, biologists have begun to identify how females partake of their own strategies: mating with more than one male, controlling (or, at least, influencing) the outcome of sperm competition, sometimes obtaining direct, personal benefits such as food or protection in return for these extra-pair copulations as well as indirect, genetic benefits that eventually accrue to their offspring. A penchant for non-monogamy among males is no great surprise, but, as we shall see, the most dramatic new findings and revised science brought about by the recent demolition of the myth of monogamy concern the role of *females*. Freud spoke more truth than he knew when he observed that female psychology was essentially a "dark continent." A well-integrated theory of female sexuality in particular still remains to be articulated; perhaps a reader of this book will be suitably inspired.

More on this, too, later. In fact, much more. It is no accident that whereas the male perspective receives one chapter, the female viewpoint gets two. It is something that we are only now beginning to identify and just barely to understand. And as is often the case with new insights, it raises more questions than it answers.

I n what follows, we shall try to keep jargon to a minimum. We've already met that deceptively simple term *monogamy*, noting the crucial distinction between social monogamy and sexual monogamy. We've also briefly considered two other mating systems: polygyny (one male mated to many females) and polyandry (one female mated to many males). The derivations of these terms make them easier to remember: *Polygyny* comes from *poly* ("many") and *gyny* ("female," the same root as "vagina"). So polygyny, the situation of a male harem-keeper, means, in effect, "many vaginas" (or cloacas, in the case of birds). Similarly, *polyandry* comes from *poly* ("many") combined with *andry* ("male," the same root as "androgen," referring to male sex hormones). So polyandry, which means "many males," is the—much rarer—situation of a female harem-keeper.

Two abbreviations will also be useful, and hence repeated throughout: EPC stands for *extra-pair copulation,* which simply means a copulation in which at least one participant is already socially mated to someone else. In human terms, it is equivalent to an extramarital affair or adultery (if the pair is married) or "cheating" (if the pair is simply "going together" or "dating" so seriously that the EPC violates the other's expectation, knowledge, or consent). For our purposes, animals can engage in EPCs no less than people can. Similarly, IPC stands for *intra-pair* (or *in-pair*) *copulation,* which means that two individuals who are socially mated also mate sexually.

Such abbreviations are intended to facilitate communication, mostly by clarity and brevity. But they have another, unintended effect, one that may be fortuitously helpful: Their aura of scientific objectivity conveys a degree of detachment that should enable us to consider difficult, emotionally charged material with at least a degree of detachment. This is probably a good thing, since not only are we as a species new to the practice of monogamy (social as well as sexual), but we are also novices when it comes to understanding this most fascinating as well as vexatious way of life.

# CHAPTER TWO

# Undermining the Myth: Males

A story is told in New Zealand about the early nineteenth-century visit of an Episcopal bishop to an isolated Maori village. Everyone was about to retire for the night after an evening of high-spirited feasting and dancing, when the village headman—wanting to show hospitality to his honored guest—called out, "A woman for the bishop." Seeing a scowl of disapproval on the prelate's face, the headman roared even louder, "*Two* women for the bishop!"

This story, of course, is one of cross-cultural misunderstanding. But underlying it is something quite different: cross-cultural, species-wide similarity, notably, the widespread male fondness for (1) sex generally and (2) sexual variety when possible. We presume that the Episcopal missionary turned down the headman's offer, but we also smile at the latter's immediate assumption that what troubled the bishop was not the prospect of spending the night with a new woman, but that he had only been offered one! To be sure, human beings can even elect celibacy (and, what is more remarkable, some can remain true to it), but virtually everyone agrees that such denial is denial indeed; it does not come naturally, and it requires saying "no" to something within.

Most people, male and female, like sex. But the Maori headman revealed an acute sensitivity to another pan-cultural human trait: the widespread eagerness of men in particular for sexual variety. This is not to claim that men necessarily seek a nonstop sexual carnival or a lifetime of wildly erotic, simultaneous encounters with multiple partners. But compared to women, men in particular—and, as we shall see, males in general—have a lower threshold for sexual excitation and a greater fondness for sexual variety, or, to look at it more negatively, a penchant for equating monogamy with monotony.

It isn't rocket science to understand how the biology of male–female differences leads to differences in sexual preference, nor is it terribly difficult to see why, from the male perspective, monogamy is so difficult. In fact, this recognition has been around for several decades now, and it has become part of the received wisdom of evolutionary biology and one of the major energizing principles of sociobiology, sometimes known as evolutionary psychology. (The other side of the coin—why the *female* perspective also generates departures from monogamy—is a different story, one that is only now being unraveled; we'll discuss it in the next two chapters).

Fundamentally, the "standard" sociobiological explanation for male–female differences is a matter of sperm and eggs. Nearly all living things are divided into male and female, and this distinction, in turn, is based on the kind of sex cells they produce, whether tiny and generated in vast numbers (sperm) or large and relatively scarce (eggs). This, in fact, is how we *define* maleness and femaleness: not by the presence or absence of beards, breasts, penises, or vaginas, and not even by who gives birth. After all, male seahorses carry their offspring inside their body, eventually releasing them to the outside after a series of violent contractions that are remarkably like those of a woman in labor. But even for seahorses, there is no question that the individual giving birth is a "he," not a "she." This is because he is the one who contributed sperm; she provided the eggs.

Most birds lack external genitals altogether, and yet biologists have no difficulty distinguishing males from females, even in those cases such as sparrows or gulls in which males and females are often indistinguishable based on physical appearance. If it lays eggs, it is female; if it makes sperm, it is male.

This is not simply a matter of theoretical dictionary definitions. It turns out that whether one is an egg-maker or a sperm-maker has important consequences. To understand these consequences, the next step is to look at the energetic expenditures that eggs and sperm entail. A female bird, for example, will lay a clutch of eggs that might well tip the scales at 20 percent of her total body weight; her male partner will ejaculate a fraction of a teaspoon of sperm. Sperm are cheap and readily replaced; eggs are expensive and hard to come by. Not surprisingly, therefore, we find that males are generally profligate with their sperm, whereas females tend to be careful and choosy about how they dispose of their eggs.

The situation for mammals is, if anything, even more asymmetric. Even though the mammalian egg is very small—almost microscopic—each sperm is smaller yet. A single human ejaculation, for example, contains about 250 million sperm, whereas, by contrast, it takes about a month for a single egg to be ovulated. (During that month, a healthy man will produce literally *billions* of sperm.) More importantly, however, each egg represents an immensely greater investment on the part of a woman—or female mammal generally—than does a sperm. If fertilized, that egg will develop within the body of its mother, nourished from her bloodstream. Following birth, the infant mammal (human or otherwise) will receive proportionately even more nourishment, in the form of milk, via the mother's breasts. By contrast, the father has only invested a few moments of his time and a squirt or two of semen, expending the energy equivalent of eating a few potato chips!

Alternatively, think about the consequences of making a mistake: If a female mammal makes a bad choice and is inseminated by an inferior male, say, one whose offspring will fail to survive or (nearly the same thing, in evolutionary terms) to reproduce, she pays a substantial toll in risk as well as in lost time and energy. Such a female may spend several weeks or many months pregnant, not to mention lactating once her offspring is born, only to have nothing to show for it on her evolutionary ledger. By contrast, a male mammal who makes himself available for one or many sexual dalliances has invested comparatively little. If in the process he succeeds in fertilizing one or more females, he is that much ahead; if he fails, then, unlike most females in a similar situation, he has not lost very much. As a result, evolutionary pressures tend to favor males who are sexually available, readily stimulated, and interested in multiple sexual relationships—who are, in the words of noted evolutionary theorist George C. Williams, "aggressive sexual advertisers." At the same time, females have generally been endowed by natural selection with a tendency to be more sexually discerning, or, as Williams has put it, "coy, comparison shoppers."

An important conceptual breakthrough came when Robert L. Trivers pointed out that the key (or, at least, one key) to male–female differences in behavior derives from differences in what he called "parental investment." Parental investment is simply anything costly—time, energy, risk—that a parent spends or endures on behalf of its offspring and that increases the chances of the offspring's being successful, at the cost of the parent's being unable to invest in other offspring at some other time. Feeding one's offspring is parental investment. So is defending, educating, cleaning, or scratching when and where it itches. And so, also, is producing the big, fat, energy-rich mother lode of nutrients called an egg. A sperm, by contrast, is a pitiful excuse for parental investment, consisting merely of some DNA with a tail at the other end.

Trivers showed that when there is a big difference between the parental investment offered by members of the two sexes, the sex investing more (nearly always female) will become a valued "resource," sought after by individuals of the sex investing less (nearly always male). Several important consequences flow from this. For one thing, males tend to compete with each other for access to females. This is because females have something of great value: their eggs or, in the case of mammals, their promise of a placenta and, eventually, lactation. Not only that, but successful males may get to inseminate numerous females, whereas unsuccessful males have nothing to show for their efforts. As a result, natural selection will generally favor males that succeed in their competition with other males and that therefore are relatively big and aggressive, outfitted with dangerous weapons (fangs, tusks, antlers, horns) and a propensity for bluff, bluster, and violence as well as sexual adventuring.

Moreover, because females generally provide such abundant parental investment, males are in many cases superfluous in terms of the success of any offspring produced. As a result, they are "liberated" to pursue as many additional reproductive opportunities as they can find. Trivers pointed out that even in cases of supposedly strict monogamy, when the direct involvement of both father and mother are required for offspring to be reared successfully, the evolutionary optimum for males will often be to adopt a "mixed reproductive strategy." In such cases, males mate with a chosen female and assist her in rearing offspring, but also make themselves available for additional reproductive liaisons with other females . . . whom, in most cases, they will not help. Because of the small investment entailed in sperm-making, males will typically be more fit, in the evolutionary sense, if at some level they are willing—even eager—to make their gametes as widely available as their lifestyles allow. It is important to realize, at the same time, that such individuals are not simply cads or scoundrels. Usually, males who seek EPCs (extra-pair copulations) are resident territorial proprietors—happily married, respectable burghers who are simply susceptible to "a little something on the side."

Once males developed this tendency to play fast and loose, it likely became self-perpetuating, so that departures from monogamy may actually be responsible, in part, for the further evolution of maleness itself; namely, for the production of especially tiny sperm. Under the pressures of sperm competition, males would probably have been pushed by natural selection to make sperm in ever-greater numbers, and—since the amount of energy that can be expended in such pursuits is ultimately limited—each one would necessarily have to be very small.

There is another way of making sense of this phenomenon of male sexual avidity, although it is not entirely distinct from looking at the low parental investment that characterize males generally. The idea is to focus, instead, on reproductive potential. In the long run, males and females have the same reproductive potential, since whenever sexual reproduction occurs, one male and one female are equally responsible. But the two sexes differ in how reproductive success is distributed among their members. Because of their high parental investment, most females are likely to be at least somewhat successful; usually there are no dramatic differences between the most successful females and the least successful. Even "low-quality" females are generally able to get inseminated, if only because males are typically ready and willing to fertilize any females who might otherwise go unmated. And because of their high parental investment, even "high-quality" females are limited in how many children they can produce. By contrast, it is possible among males for a small number of well-endowed individuals to be hugely successful, while others are total failures.

Consider this: During her lengthy pregnancy, a cow elk is fully occupied with only one calf. By contrast, a bull can inseminate additional females every day. Admittedly, the fact that most healthy male mammals—including mice and men—release a few hundred million sperm in just one ejaculation does not mean that they are capable of fathering a few hundred million children. Take the human situation, however: During the nine months that a woman is pregnant—not even counting the added time lactating—a man has much greater reproductive potential . . . that is, *if* he inseminates additional women. Another way of looking at it: What limits the reproductive success of any given male would appear to be his access to females, rather than inherent limitations of his reproductive anatomy.

And so, once again we have the same basic pattern: Males, which make a relatively small parental investment and have a large potential reproductive success, tend to be sexually eager. This does not in itself require departing from monogamy, but in fact such sexual eagerness is likely to be especially pronounced when it comes to new potential partners, who, once inseminated, will produce offspring yielding the kind of evolutionary payoff that selects for precisely such behavior. By contrast, a rigidly monogamous male—with no eye for the ladies other than his own—has fewer opportunities for reproductive success. The result? From the male perspective, strict monogamy is not likely to be the best of all possible worlds.

It is said that exceptions prove the rule. When it comes to the connection among maleness, low parental investment, and sexual eagerness, there are in fact some interesting apparent exceptions. These are cases of "reversed sex roles," in which females are comparatively aggressive, often larger, brightly colored, and more sexually demanding if not promiscuous, while the males are coy, drab, and sexually reticent. Among certain insects, for example, the males produce not only sperm but also a large mass of gelatinous, proteinaceous glop, which the female devours after mating; in doing so, she gains substantial calories, more, in some cases, than she expends in making eggs. And sure enough, in these species (including some katydids and butterflies), females court the males. This makes sense, since here it is the males, not the females, who make a large metabolic investment. And in such cases, males, not females, are likely to say "no." The key for our purposes—and apparently for these animals as well—is that male–female patterns of sexual behavior are reversed precisely when male–female patterns of parental investment are reversed. (It is not known, incidentally, what gave rise to such sex-role switching in the first place.)

Another lovely insect example of the exception confirming the rule comes from several species of fruit fly, including one known as *Drosophila bifurca*. The males, about 2.5 millimeters long, make sperm that are 20 times longer than the flies producing them! This would be the equivalent of a 6-foot man making sperm that are more than 120 feet long. The function of these giant *Drosophila* sperm is unknown (in some cases, the tail enters the egg at fertilization, in others it remains outside), but it is known that males, after constructing these remarkable devices, parcel them out "with female-like judiciousness, carefully partitioning their limited sperm among successive females."

Sex-role reversal has even been reported for some species of birds, notably the South American marsh-dwelling species known as jacanas. These animals are polyandrous, with a large, aggressive, dominant female maintaining a territory in which several small, meek, and subordinate males each construct a nest and incubate eggs that the dominant female bestows upon them, after mating. Because of their time spent nest-building and incubating, the male jacanas end up providing more parental investment than do the females . . . and once again, the females act "male-like" in their sexual appetites, while the males behave more like females.

Of course, exceptions don't really prove rules; rather, *apparent* exceptions can help contribute to a rule if a more careful look shows that they aren't really exceptions after all. (Otherwise, exceptions *disprove* rules!) But when it comes to the correlation between low parental investment and high sexual appetite, this rule is pretty close to being proven.

By now, it should be easy to see why monogamy is under siege, at least from the male side. The potential reproductive benefit of having one or more additional sexual partners is high (if any of these "girlfriends" get pregnant), while metabolic and energy cost is likely to be low. Not surprisingly, males show numerous signs of this evolutionary pressure to stray from monogamy. One example is the so-called Coolidge effect.

Legend has it that President Cal and his wife were separately touring a model farm. When the president reached the chicken-yard, containing a single rooster and several dozen hens, his guide said, "Mrs. Coolidge wanted me to point out to you that this one rooster must copulate many times per day." "Always with the same hen?" asked Coolidge. "No, sir," replied the guide. "Please point out *that* to Mrs. Coolidge!" the president responded.

The Coolidge effect is well known and has been confirmed in numerous laboratory studies: introduce, for example, a ram and a sexually receptive ewe and the two will likely copulate, typically more than once. The frequency then declines, usually quite rapidly. But replace the female with a new one, and the seemingly "spent" ram is—to some extent—sexually reinvigorated. A new ewe makes for a new him.

Actually, this phenomenon was known long before the modern science of animal behavior. "I have put to stud an old horse who could not be controlled at the scent of mares," wrote the sixteenth-century essayist Montaigne. "Facility presently sated him toward his own mares: But toward strange ones, and the first one that passes by his pasture, he returns to his importunate neighings and his furious heats, as before."

As to human beings, listen to this account by a man from the African Kgatla tribe, describing his feelings about sexual intercourse with his two wives:

> I find them both equally desirable, but when I have slept with one for three days, by the fourth day she has wearied me, and when I go to the other I find that I have greater passion; she seems more attractive than the first. But it is not really so, for when I return to the latter again there is the same renewed passion.

There is no reason to think that men inhabiting modern technological societies are any different. Indeed, the famous team of sex researchers led by Dr. Alfred Kinsey pointed out that

> most males can immediately understand why most males want extramarital coitus. Although many of them refrain from engaging in such activity because they consider it morally unacceptable or socially undesirable, even such abstinent individuals can usually understand that sexual variety, new situations, and new partners might provide satisfactions which are no longer found in coitus which has been confined for some period of years to a single sexual partner. . . . On the other hand, many females find it difficult to understand why any male who is happily married should want to have coitus with any female other than his wife.

About 80 percent of all mammal species are capable of multiple ejaculations, an ability that makes particular sense if these multiple ejaculations involve multiple sexual partners. In addition, although the Coolidge effect is very widespread, non-monogamous species (primates as well as rodents) show a stronger Coolidge effect than do monogamous species; this also was predicted, since males of non-monogamous species have more opportunity to act upon any sexual excitation they experience when they encounter a new female.

Most observers of animal behavior, not to mention observers of *Homo sapiens,* would agree that males generally have greater sexual urgency and lesser discrimination: Ask yourself, for example, is it men or women, who are accused of date rape, who engage in various sexual paraphilias ("perversions"), who visit prostitutes, and who have made pornography one of the largest industries worldwide? Once again, it makes biological sense for the sex that produces cheap, easily replaceable gametes to be readily "turned on" sexually and to be comparatively undiscriminating as to the target. (Thinking about it objectively, and without an evolutionary perspective, it is rather bizarre that huge numbers of men find it highly arousing to look at visual images of naked women! After all, these are merely dots of color on a page or, increasingly, a computer screen. Such people are not morons; intellectually, they know that these arousing images are just that—images—but the male tendency is to have an especially low threshold for sexual stimulation.)

What does an individual "get" from an EPC? For males, at least, it is obvious: sexual satisfaction. But this is what biologists call a "proximate" explanation. It may explain the immediate causation, but it leaves unanswered the deeper question: Why should sex—especially, perhaps, sex with someone new—be gratifying? The answer, for evolutionary biologists, is also obvious: Something is gratifying if it serves the biological interests of the individuals concerned. Such proximate satisfactions as "gratification" are evolution's way of getting creatures to do certain things. Or, more accurately, those who find such activities rewarding, and who therefore engage in them, leave more descendants who thus have similar inclinations. Hence, we find animals who seek food when hungry, rest when tired, warmth when cold . . . and sex when horny.

Looking now at the evolutionary payoff of EPCs, it seems that males are unlikely to gain anything other than an increase in their reproductive success. After all, they have to expend time and energy seeking EPCs, and they may also be attacked by an outraged "husband." Moreover, since they are the eager ones and are providing only sperm—which, after all, are cheap—it is unlikely that their female EPC partner will lavish "gifts" upon them, in the form of extra food, donated territory, assistance in defending their own offspring, and so forth. Whereas a mistress may gain material rewards from her lover (whether animal or human) in return for her sexual favors, a male out-of-pair sexual partner rarely is "paid" by his inamorata. For the male Lothario, the rewards are more likely to be immediate (the gratification of sexual dalliance itself) and long term (enhanced reproductive success), rather than material. For females, as we shall see in Chapters 3 and 4, the situation is quite different.

In any event, it requires no great conceptual leap to see how the low cost of sperm (and the resulting potential for a high reproductive rate) leads to a low threshold for sexual stimulation as well as a predisposition for multiple sexual partners; nor is it difficult to see how these, in turn, lead to a penchant for polygyny or, in cases of monogamy, to a susceptibility to EPCs. Either way, male biology bodes ill for monogamy. Interestingly, the biology of mammals is even more stacked against monogamy. This is because among birds, nestlings often have very high metabolic needs and therefore require the efforts of two committed parents. As a result, although male birds can often be expected to seek EPCs, they are somewhat less likely to maintain a harem of females. Not that male birds wouldn't happily attempt to accommodate such an arrangement; rather, their needy offspring are generally so demanding that most males are constrained to pitch in and, therefore, to have only a limited number of mates. But female mammals are uniquely equipped to nourish their offspring; indeed, mammary glands are what sets mammals apart from other animals. As a result, we can expect mammals to be even more predisposed than birds to form harems; that is, to be polygynous. And in fact they are. We already mentioned that monogamy is very rare among mammals. But we also mentioned the recent, dramatic discovery that even birds—including those that are socially monogamous—are much more prone to EPCs than anyone had imagined.

From an evolutionary perspective, copulations themselves don't really count; fertilizations do. And EPCs can be highly effective for males, whether bird, mammal, or any other species. Among red-winged blackbirds, more than 20 percent of a male's reproductive success comes from EPCs. Also, males having a high reproductive success with their own mates are likely to have high reproductive success via EPCs; in the world of reproduction, the rich get richer.

In fact, males who succeed in obtaining EPCs are in most cases already mated. This makes sense if the same underlying desirability of certain males that renders them more likely to obtain mates in the first place also contributes to their success when they go outside their mateships, seeking EPCs. But even here, there are exceptions. DNA profiling has recently allowed researchers to identify all the individuals, as well as all the offspring, of a small population of "stitchbirds," which are seemingly unremarkable little songsters except for this peculiarity: They copulate face-to-face, a position that may well be unique among birds. These animals were studied on an island off the coast of New Zealand, more precisely, Tiritiri Matangi. (For those readers who are geographically challenged, this will doubtless help: Tiritiri Matangi Island is about 3 kilometers off the Whangaparaoa Peninsula. Enough said.)

During the two breeding seasons in which the stitchbirds were studied, there was a heavy bias toward males in the population, a ratio of 3 to 1 in one year and 2 to 1 in the next. Nonetheless, social monogamy is the apparent goal of all self-respecting stitchbirds. But as a result of the unbalanced sex ratio, there were a large number of seemingly unsuccessful, bachelor males, so-called floaters. Thanks to a penchant for EPCs, however, these floaters were in no way reproductive losers: More than one-third of all nestlings were fathered by floaters (for another way of looking at it, 80 percent of all nests contained at least one extra-pair nestling). In fact, two of the unpaired floaters achieved more fertilizations than at least one paired male who played by the rules; this male had succeeded in pairing with a female in both years of the study, yet because of EPCs, his reproductive payoff was less than that of the two excluded floaters.

With findings like these, the importance of EPCs has become undeniable. It is also tempting to go further and conclude that males engaging in EPCs somehow experience a reproductive advantage over their IPC (in-pair copulation) counterparts. Accordingly, British biologist Tim Birkhead and his colleagues conducted an experiment to determine the effectiveness of EPCs in achieving fertilization. The subjects were zebra finches: small, brightly colored, socially monogamous birds native to Australia and commonly kept as pets throughout the world. In the wild, males guard their females before and during egg-laying, copulating about twelve times for each clutch. After allowing mated females to copulate on average nine times with their mated male, the experimental subjects were exposed to another male and allowed to copulate once. Both the mated female and the extra-pair male obliged. When the researchers then used genetic plumage markers to determine paternity, they found, on average, that just one EPC yielded 54 percent of the offspring, compared to 46 percent paternity resulting from nine IPCs!

Much of this success—perhaps all of it—is due to a very strong "last male advantage," especially pronounced in many birds: The last male to copulate with a given female before she lays her eggs enjoys disproportionate success in fertilizing those eggs. Regardless of the mechanism, however, the key point for our purposes is that a small amount of breeding effort on the part of males can yield dramatic returns. ("Last male advantage" also has important consequences for the behavior of females, as we'll see in the next chapter.)

Many animal breeding systems are such that a small number of suc-cessful males are able to monopolize—at least socially—a larger number of females. In such cases, EPCs could, in theory, level the reproductive play-ing field if otherwise excluded bachelor males gain some "sneak fertiliza-tions" while the legitimate husbands aren't watching. But, instead, EPCs usually *increase* the differences among males. To be sure, it is now a com-monplace that some of the offspring previously attributed to a given male (the "husband," if socially monogamous, or the harem-keeper, if polygy-nous) are likely to have been fathered by someone else. But at the same time, success in achieving EPCs is not randomly spread among males; just as certain males are more successful than others in obtaining social mates, certain males are also more successful than others in obtaining extra-pair copulations and, thus, genuine paternity. Nearly always, it is an already successful husband or harem-keeper who also gets the EPCs. For example, among European red deer (called elk in the United States), it has long been known that harem-masters are remarkably successful, whereas excluded bachelors are, comparatively speaking, evolutionary losers. Now, DNA fin-gerprinting shows that the actual difference between genetic "haves" and "have-nots" is even greater than what had earlier been estimated based on behavioral evidence alone.

Part of the reason for the success of EPC-ing males may lie in this observed fact: Compared with a female's regular sexual partner, they actu-ally produce more sperm at a given copulation, especially if they have not mated recently. In one study involving birds, this was measured by the ingenious if indelicate procedure of persuading males who had been ab-stemious for different time periods to copulate with a freeze-dried female who had been fitted with a false cloaca.

We don't know whether sperm numbers tend to be generally higher dur-ing EPCs than IPCs for mammals as well, nor whether this holds for human beings. We do know, however, that men generally report a higher level of sexual excitement with a new partner (remember the Coolidge effect). Since IPCs, by definition, cannot be with a new partner—except once—it is at least possible that EPCs involve, on average, the production of more sperm per ejaculation among human beings, too. If so, it would make human EPCs more likely to result in offspring than chance alone would predict. Needless to say, this speculation will be difficult to test . . . but not impossible.

It is a well-established principle that among polygynous, harem-keeping animals, males are larger than females. Compare, for example, gorillas with gibbons. Gorillas establish harems, in which a dominant "silver-back" male mates with perhaps three to six females; male gorillas are two to three times larger than their mates, apparently because competition among males to be successful harem-keepers has conveyed an evolutionary advantage to those that are larger, stronger, and generally more effective in keeping their rivals at bay. By contrast, gibbons live in male–female pairs, so most healthy gibbons get to reproduce, even most males. As a result, there are very few big winners or big losers and, accordingly, virtually no size difference between the sexes.

Among monogamous species generally, males and females ostensibly have the same reproductive options. After all, when the female reproduces, so does the male, and vice versa. Such animals therefore shouldn't be sexually *dimorphic* (from the Greek for "two bodies"), because natural selection ought not to reward either sex for being overgrown, remarkably colored, or otherwise extravagant. But many animals long known to be socially monogamous are in fact sexually dimorphic, with males typically being more brightly colored than females—especially among birds. Consider mallard ducks, for example, in which the drakes have a dramatic, iridescent green head, whereas females are comparatively drab, or the many species of warblers in which, once again, males are extraordinary in their bright coloration, whereas the females are notoriously difficult to tell apart.

Darwin thought that perhaps dramatic male–female differences were maintained in such cases because more elaborately ornamented males got to breed earlier; as a rule, the early bird not only gets the worm, but he— or she—also gets to have more successful offspring. So, one possible explanation for fancy secondary sexual traits among males in monogamous species is that for one reason or another, sexier males get to mate with more fecund females. A second possibility, raised only recently as the myth of monogamy bites the dust, is that sexier males are able to gain additional reproductive success through EPCs, at the expense of mated males that aren't as attractively adorned. Over time, this would increase the proportion of fancy-looking males or, at least, of males that are quite different from females of the same species. (This leads to an interesting, if troubling, possibility: As we shall see in Chapter 5, a powerful piece of evidence for the fact that human beings are biologically polygynous is the fact that men are generally larger than women. Although the evidence is still convincing in that direction, it is also possible that our own biological history was largely monogamous, with at least some male–female differences in *Homo sapiens* due to the fact that adultery figured importantly in that history.)

Back to the birds, where an important recent study focused on a small species known as the collared flycatcher, on the Swedish island of Gotland. It provides strong evidence that extra-pair copulations in this supposedly monogamous animal give certain males a distinct reproductive advantage. Collared flycatcher males have a white forehead patch; females don't. This patch is a secondary sexual characteristic. It also appears to be a status signal, whose size varies depending on its bearer's nutritional status as well as social success. Knowing the size of this patch, we can predict which collared flycatcher will win a territorial dispute between males. Moreover, if experimenters artificially increase the size of this patch, the fortunate males are more likely to establish a territory in the first place. Female collared flycatchers mated to males with large white forehead patches even produce a larger proportion of sons. This seemingly odd finding makes sense if males with prominent patches are likely to father sons with prominent patches. There would then be a reproductive payoff if such males father a comparatively large number of sons, which, in turn, will be relatively more successful than daughters . . . who lack such distinguishing marks and, as females generally, are less likely to distinguish themselves reproductively.

The researchers found extra-pair paternity among collared flycatchers in 26 of 79 broods, accounting for 71 of 459 nestlings. After carefully analyzing their results, they concluded that "selection via variation in paternity"—that is, some males having offspring via EPCs with females mated to other males—can be more important than "selection resulting from mate fecundity" (that is, more important than having greater numbers of offspring with their socially defined mates). Among collared flycatchers, in short, the major route to male reproductive distinction is not having more offspring with their "wives" but rather fathering offspring by various "lovers" who are already "married."

There have been many other studies along these lines, looking at species that are socially monogamous, most of them showing not only that males with highly developed secondary sexual traits have higher reproductive success but also that such success comes via EPCs. A typical one, in the prestigious journal *Nature*, was titled "Extra-Pair Paternity Results from Female Preference for High-Quality Males in the Blue Tit." Other studies have shown that male paternity—the proportion of offspring in his nest that are genetically his, as well as the number of offspring he will father in the nests of other males—is connected not only to the presence or absence of certain secondary sexual traits but also to the degree to which his secondary sexual characteristics are expressed.

Here are some notable examples: "DNA Fingerprinting Reveals Relation Between Tail Ornaments and Cuckoldry in Barn Swallows" showed that male barn swallows sporting more deeply forked tails are more likely to win the hearts of neighboring females. Another article, "Correlation Between Male Song Repertoire, Extra-Pair Paternity and Offspring Survival in the Great Reed Warbler," reported that in this European species—like barn swallows, ostensibly monogamous—males with a large variety of songs are also likely to have a variety of sexual partners. And don't overlook this gem, potentially encouraging to any readers getting a bit long in the tooth: "Old, Colorful Male Yellowhammers, *Emberiza citrinella,* Benefit from Extra-Pair Copulations." Among these birds, males grow more colorful as they age. It appears that older, more colorful males therefore give promise of having a desirable set of longevity genes, which in turn are attractive to females. (Human beings, too, grow more colorful as they get older, and it is at least possible—if unlikely—that white hair has been selected as a similar symbol of status or, at least, of the ability to survive.)

It seems that the key dimension in these cases is female choice; since males of most species are unlikely to refuse a quick and easy EPC, the deciding vote as to who succeeds and who fails is generally cast by the females, based on whom they find most attractive. But the secondary sexual characteristics of males not only dictate their attractiveness to females, they also influence—and are influenced by—dominance relationships among males. And so, the two factors—dominance relationships and degree of secondary sexual traits—are confounded when it comes to determining which males obtain EPCs. Among cattle egrets, for example, the dominance standing among males has implications for who gets to have EPCs with whose mate. Dominant males have EPCs with the wives of subordinates but not vice versa. (This pattern is of course not unknown among *Homo sapiens,* too.)

On the other hand, one thing about biology—as compared with, say, chemistry or physics—is that there are lots of exceptions. This applies to the general correlation between male secondary sexual traits and reproductive success no less than to other generalizations (such as "animals eat plants but not vice versa," "only mammals are warm blooded," or "females are smaller than males"). There are, after all, insectivorous plants; it appears that dinosaurs were warm blooded, and among some species—such as jacanas, described earlier—females are larger than males. Similarly, male secondary sexual traits don't always correlate with reproductive success; that is, sexy males don't always get more EPCs, or—more important—they don't always get more EPFs (extra-pair fertilizations). Are these, like the examples of sex-role reversal described earlier, cases of exceptions "proving" the rule? It's too early to say.

I t is clear that even in ostensibly monogamous species, males seek—and often obtain—EPCs. It is also clear that they do so at the expense of other males, namely, the ones "married" to those females who succumb to their charms. For an EPC-seeking male, the best arrangement is to father children with females who are already mated. In such cases (assuming the female is able to deceive her in-pair mate as to her infidelity and, thus, his nonpaternity), the female will gain paternal assistance from the cuckolded male, making it more likely that any offspring thereby conceived will be successful. At the same time, the extra-pair male receives this additional payoff: Since any resulting offspring will be reared by someone else, there is no additional parental effort required. Not only that but—at least in the case of birds—by having multiple sexual partners, extra-pair males literally succeed in placing their eggs in more than one basket. On the other hand, an EPC with an already-mated female is almost always riskier than one with an unmated female, since the outraged husband may find out and drive off an interloping male, possibly injuring him.

For the potential cuckold, therefore, an important option is to guard "his" female from gallivanting males, thereby possibly preventing them from achieving their aims.

(All this assumes, by the way, that the result of a successful EPC is that a mated female ends up bearing offspring sired by one or more extra-pair males and that, therefore, the in-pair male is the one who is victimized. But occasionally the in-pair female can be the loser: In one recorded instance, a mated male zebra finch succeeded in inseminating an unmated female, who then laid an egg in that male's nest. The result was one of the few documented cases in which, as a result of an EPC, the cuckolded party was a *female*—in whose nest the egg was laid—rather than a male. Most of the time, however, cuckolds are male, and for good reason: Among all species with internal fertilization—including birds, mammals, and reptiles—a female "knows" that any offspring emerging from her body is genetically hers, whereas a male has to take his mate's word for it, unless he is an especially assiduous mate-guarder.)

Let's take a brief excursion to a cattle pasture, almost anywhere in the world. One common occupant—far more numerous than the cows—is a tiny insect, the yellow dungfly. Male dungflies gather around cowpatties, especially the fresh ones, where they search for females who are about to lay their eggs in the warm, gooey interiors. Interestingly, these arriving females have nearly always copulated *before* arriving at the egg-deposition sites. Therefore, they already contain enough sperm in their genital storage organs to fertilize all their eggs. Yet before ovipositing, they copulate once again with at least one waiting male. Why?

The answer seems to be that among dungflies, the last male to mate guards "his" female until she is finished laying her eggs. Having such a protector greatly reduces the harassment that a female would otherwise receive from other males. This isn't a bad deal for males, either, since they experience a "last male advantage": The last male to mate fertilizes more than 80 percent of a female's eggs. Guarding takes about a quarter of an hour, which is probably a good trade-off.

Guarding doesn't always work, though. If another male dungfly, much larger than the guarder, attacks the pair, he may succeed in copulating yet again with the female, after which he proceeds to stand guard himself.

Mate-guarding is very widespread. It is even possible that the well-known tendencies for animals of many different species to establish and defend territories is really a consequence of males guarding the sexual rights to their mates by defending a region surrounding them. In a sense, mate-guarding is one of the clearest animal (or human) manifestations of sexual jealousy, and it is sometimes quite overt, with the male shadowing every movement of his mate. Such "togetherness" is almost certainly not a simple—or even complex!—matter of love or loneliness, since it is nearly always limited to precise times when the female in question is fertile. Male bank swallows, for example, closely follow their mates, flying along nearby whenever they venture from their nests; such devoted attention quickly terminates, however, when the females are no longer fertile.

Mate-guarding is also a common male strategy among mammals, especially when the female is in estrus. The goal, once again, is evidently to thwart EPCs. Mate-guarding is also widespread, almost universal, among that mammalian species known as *Homo sapiens:* A now-classic anthropological review recorded that only 4 out of 849 human societies did not show some signs of mate-guarding, whereby men keep close tabs on their mates. In some societies, husbands even time their wives' absences while they are in the bushes urinating or defecating. Such concern may not be ill founded; one piece of British research found that the less time a woman spent with her primary mate (husband or identified main sexual partner), the more likely she was to have copulated with someone else.

We used to think that the close association of male and female, especially in a monogamous species, was simply a manifestation of their close pair-bond, as well as perhaps a way of further enhancing their relationship. Here is a description of courtship among European blackbirds, from a classic 1933 account by ornithologist E. Selous:

> The male bird follows her all about, hopping where she hops, prying where she pries, and seeming to make a point of doing all she does except actually collect material for the nest. . . . Then, the one laden, the other empty-billed, they both fly back in just the same way, and the cock will sit again . . . for the cock is as busy in escorting and observing the hen as she is in collecting material for building the nest.

Today: same observation, different interpretation. The male's observing and escorting seem motivated less by love or chivalry than by sexual jealousy and the specter of EPCs. The relationship of mate-guarding to EPCs is complex, and expected to be. On the one hand, we might anticipate a negative relationship: the more mate-guarding, the fewer EPCs. This seems the most obvious connection. But on the other hand, if EPCs are rare occurrences in a particular species, then we wouldn't expect males to waste a lot of their time defending against such a nonexistent threat. There is in fact quite a range out there in the natural world, from essentially no correlation between mate-guarding and EPCs, to a positive association, to a negative one. (The determining factor may well be where each population is being sampled in its evolutionary history: If there is a continuing arms race between males seeking EPCs and those seeking to prevent them, then it might see-saw back and forth, with one side or the other ahead in different species at different times.)

As we now understand it, mate-guarding can take many forms, with—not surprisingly—some of the more bizarre patterns revealed by insects. Among numerous species of bees, for example, the lower abdomen of the male almost literally explodes after mating, part of it then adhering to the female and thereby providing a kind of posthumous mate-guarding. Among a group of insects known as phasmids, genital contact during mating can last as long as 79 days; this tantric excess can be seen as an extreme example of mate guarding.

The widespread male preoccupation with mate-guarding fits our expectations of something motivated by the threat of EPCs. Thus, it is typically more intense with increased risk of the female's engaging in an EPC. Among barn swallows, which nest both solitarily and in colonies, females are guarded more closely in the latter case (when the proximity of numerous males makes EPCs more likely) than in the former (when there are no sexual competitors nearby).

On the other hand, for all its logic, mate-guarding presents us—and the mate-guarders—with an interesting irony. It wouldn't be needed if males themselves were not gallivanting about, seeking EPCs. If no one gallivanted, no one would have to mate-guard. Moreover, mate-guarders themselves, when not on guard, are likely to be out gallivanting! In fact, it may well be that the major constraint against doing yet more gallivanting is that while off seeking EPCs, a gallivanting male may find himself cuckolded . . . by another gallivanting male!

Recall those male bank swallows who so attentively fly after their females. Male bank swallows are socially monogamous and assist their mates in building the nest and incubating and feeding the young. In addition, they regularly seek EPCs with other females, before and after pair-bonding. Accordingly, they gallivant about in search of EPCs and also guard their mates against other gallivanting males, but, of course, they can't do both simultaneously. For seven to nine days after pair-formation, the male pursues the female whenever she flies from her nest burrow—up to 100 times per day. Other males seek to make contact with the female during these flights; in fact, the labored, heavy flight of a female carrying unlaid eggs may serve as a cue to the other gallivanters. For his part, the guarding male actively seeks to drive the female back to the home burrow, especially when she flies into heavy traffic and may be pursued by three or four bank swallow males, starry-eyed with visions of a quick EPC.

For about four days immediately prior to egg-laying, when copulations lead to fertilizations, the male bank swallow is very busy, attentively guarding his female. Before this time, as well as after—that is, when her eggs are not ripe, and again after his genes are safely tucked away inside the shells—he goes seeking EPCs with the mates of other males . . . who, of course, are busy with defensive mate-guarding of their own.

It is unlikely that these chases are "sexual displays," intended to enhance the pair-bond, as earlier literature in animal behavior had suggested. This is because (1) males always chase females, not vice versa; (2) males typically fight with other males as an immediate result of such chases; and (3) when their own female is no longer fertile, mated males typically join in chases of other females, even though bank swallows are strictly monogamous, at least at the social level. Thus, such males could not be solidifying an additional pair-bond, if only because no such "double-bonded" males have ever been found.

David has traced a comparable pattern of mate-guarding and gallivanting among hoary marmots, social-living, mountain-dwelling relatives of the eastern woodchuck. The male makes periodic forays beyond his colony area, apparently in search of EPCs with fertilizable females. These episodes are significantly more frequent early in the season, when females are in estrus. Alternatively, a male marmot sometimes remains close to his female, guarding her from other sexually motivated males. It is clear that this passion for togetherness is initiated by the male, not the female, since during times of guarding, physical proximity is maintained by his movements, not hers. Furthermore, males are more likely to gallivant when their female is within her burrow, rather than out in the mountain meadows, and also when their neighbors are predominantly adult females rather than adult males. When the opposite condition holds—lots of other males nearby—mated males, not surprisingly, concentrate on guarding. Females of this species breed in alternate years; as expected, males that are associated with nonbreeding females go gallivanting, whereas during the season that his female is reproductive, the male stays at home, mate-guarding.

The evolutionary benefit of mate-guarding depends on how many other males are doing the same thing: If everyone else is staying home and mate-guarding, a would-be gallivanter wouldn't have to worry that while he was out seeking EPCs, he might be cuckolded by other EPC-seeking males. But at the same time, the more mate-guarding he does, the less likely our male is to achieve his sought-for EPC. Lots of other gallivanters means a greater chance of being cuckolded, but also a greater chance of gaining access to someone else's unguarded females. No one said that these things were going to be easy! The best way to understand such complex trade-offs is through the mathematics of game theory, which is concerned with examining interactions for which the payoff depends on what other "players" are doing. Although this is not the place to develop such an analysis, it is a good place to point out that, in a very real sense, the dilemma of EPCs versus mate-guarding is one of the males' own making (with a little help from the females).

When males are mate-guarding, they rarely gallivant. This suggests that the former has priority over the latter, which makes sense since the prospects are usually better of preventing an outsider from copulating with your own mate than of obtaining an EPC with someone else's. But when that mate is no longer fertile, all bets are off; at this point, males typically switch to gallivanting. For example, consider rock ptarmigans, partridge-like birds of the arctic-alpine regions. When the females are fertile, males average one intrusion onto a neighbor's territory every 14 hours. As soon as their mates are infertile, however, male intrusion rate jumps to one intrusion about every 1.4 hours, a 10-fold increase. And you can bet that those intrusions are not simply intended to exchange pleasantries or talk about the weather.

A new cottage industry has sprouted among field biologists. In addition to using DNA fingerprinting to find out whether the "husband" of a seemingly monogamous pair is also the father, interest in EPCs and mate-guarding has given rise to a slew of research studies in which males are live-trapped and kept away from their mates for various periods of time. The intent is to see whether extra-pair males take the opportunity to "make a move" on the temporarily abandoned females. If so, this suggests that mate-guarding is normally important in the species. Just about always, it is.

For example, when male wheatears (small, monogamous birds) were removed for 24 hours during the female's fertile period, the frequency of intrusions by neighboring males and the number of EPCs shot up. By contrast, when males were removed during incubation (a time when females are no longer capable of being fertilized), there was no such increase. Extra-pair males are evidently able to determine whether a temporarily single female is likely to be sexually receptive. Not surprisingly, extra-pair paternity was higher (by about 25 percent) when the paired male was experimentally removed.

What characterized successful intruding males? In addition to the secondary sexual traits already mentioned, the key seemed to be the body condition (weight, overall health) of the intruders relative to the mated males that had been removed. Extra-pair males that were in poorer condition than the removed males were resisted by the females, who, it should be noted, continued to remain with their nest despite being made (albeit temporarily) into single parents. Similarly, extra-pair males that succeeded in copulating with the "visited" female were always in better body condition than the removed male. Not surprisingly, therefore, experimentally removed males that had been in poor body condition were more likely to wind up with "offspring" not their own. This suggests that the female has a lot to say about whether EPCs take place and, if they do, whether they are successful in actually fertilizing her eggs. (As we shall see, it also suggests what females are looking for when they engage in EPCs.)

It may well be that it is largely the females, not the males, who ultimately control the paternity of their offspring, although it seems that mate-guarding is an important factor. (If it weren't, then presumably males wouldn't bother doing it!) Among wheatears, at least, females were never seen behaving territorially toward intruding males; they allowed themselves to be visited and often courted, but they did not necessarily copulate with their extra-pair suitors.

Males can be extraordinarily crafty in setting up EPC opportunities for themselves, taking advantage of the preferences of females and, one way or another, getting around any efforts at mate-guarding on the part of in-pair males. Think about this series of events, observed among a species of gibbons living freely in the rain forests of Southeast Asia. Subadult males are tolerated within the family group, and perhaps not just because of paternal benevolence: In one case, while the male of a gibbon group was involved in an encounter with a subadult from another group, the adult male from the subadult's group rushed into the adjoining territory and achieved a copulation with the female! The possibility arises that adult males tolerate subadults in their group because the subadults occasionally get adjacent males socially entangled, giving the adult males an EPC opportunity.

More manipulating: In a series of observations of purple martins (birds related to swallows), it was found that older males monopolized several nestboxes and mated monogamously, one female to each male. Later, after that female was incubating—and thus was no longer fertile—the older males vocalized and thereby seemed to attract a following: Younger males set up housekeeping in the adjacent boxes. The older males then proceeded to obtain EPCs with the nubile mates of these less experienced, and presumably less attractive, young adults. As a result, the younger males fathered a mere 29 percent of the eggs *in their own nests*, whereas the older Casanovas produced, on average, 4.1 offspring via their female partner plus an additional 3.6 offspring via the mates of neighboring males.

(There may be a human parallel here, notably those charismatic men who establish cults or other forms of communal living arrangements and then proceed to monopolize the sexual attentions of the women, including those ostensibly associated with other, more junior cult members. Indeed, one of the main reasons for the failure of various utopian communes has been eventual resistance to the sexual privileges typically demanded—and received—by the founding fathers.)

Males that are cuckolded are in double jeopardy: Not only are they more at risk of losing out genetically to gallivanters, but they are also less likely to be successful themselves in seeking their own EPCs. Why? Probably because those males especially likely to be cuckolded suffer this indignity because of some shortcoming in themselves. So whatever inclines their mates to seek matings elsewhere is also apt to make those same males unappealing to other females. They are losers two times over.

As a general rule, since the females of high-quality males are less likely to engage in EPCs, high-quality males have less need to guard. Older, more attractive males thus have a double advantage over their younger counterparts: Not only are they evidently appealing as sexual partners to already-mated females, but because they are so desirable, their own females are less likely to engage in EPCs. So these males have little need to spend time and effort mate-guarding and are therefore freed up to seek EPCs. The general pattern is concisely described in the title of one research article: "Unattractive Males Guard Their Mates More Closely." Several studies have confirmed that poor-quality males are generally more concerned with mate-guarding than are their high-quality counterparts, and for good reason, since females whose mates are less desirable are more inclined to seek EPCs. Males in poor condition are the Avis of Aves: They try harder.

And not only birds: It must be noted that among human beings, less attractive men invest more time and money in their mates than do men who are more attractive.

Aside from mate-guarding, how else can mated males diminish the threat posed by EPCs? In some cases, males have other anti-EPC tactics up their sleeves. For example, male swallows returning to their nests and finding their female absent typically give a loud alarm call, which causes all the birds in the colony to fly up in excitement. Several times this was seen to disrupt an EPC that the alarm caller's mate was engaged in. Maybe the husband was truly alarmed that his wife wasn't home. In any event, mated male swallows who experience this special kind of empty-nest syndrome are particularly likely to give alarm calls if they inhabit a crowded colony; faced with the same circumstances, solitary householders usually keep quiet.

Males can also fit their mates with the equivalent of a chastity belt, a "copulatory plug." Among many species—including most mammals—part of the seminal fluid coagulates and forms a rubbery mass that is often visible, protruding slightly from the vagina. It used to be thought that these copulatory plugs served to prevent sperm from leaking out. And well they might. But it is increasingly clear that they also work the other way: to keep other males from getting in. It warrants repeating: Such devices would not be necessary if females weren't inclined to mate with more than one male.

In the world of spiders, males are often attracted by female pheromones, which waft downwind from their web. Not uncommonly, however, males will destroy a female's web after mating with her. Although not a chastity belt, such actions represents something similar: the male's effort to inhibit his mate's sexual activity. By ruining her web, the male drastically reduces the chances that another male will find and mate with the same female.

Please don't get the impression, incidentally, that females are merely passive bystanders or victims of all this sexual skulduggery. After all, in this chapter we are intentionally focusing on male strategies; in the next two, we concentrate on females. Biologist William Eberhard has been especially influential in pointing out the likely importance of what he calls "cryptic female choice," whereby females select which sperm will receive favored treatment and be admitted to their precious eggs. But insofar as females are exercising such choice, it is likely that males will try to horn into the act, each attempting to bias the outcome in his favor. Eberhard reviewed the sweaty details of what male insects actually do during courtship and copulation in 131 different species; he found that 81 percent showed behavior during copulation that he considered to be "copulatory courtship," activities that go beyond the simple necessities of transferring sperm and that appear to be directed toward persuading females to transfer and retain *their* sperm, in preference to the sexual products of other males.

Another possible male response to the EPC threat is frequent copulation. In mate-guarding, a male uses his body to keep other males at bay; by relying on frequent copulation, he uses his sperm . . . lots of them, delivered often. The idea is simply to overwhelm the opposition, to swamp their sperm with one's own.

Of the two tactics, it seems likely that mate-guarding is more efficient; after all, successful mate-guarding means there is essentially no chance of being cuckolded, whereas frequent copulation simply invokes probability, attempting to tip the genetic scales in one's favor. Also, even though sperm are cheap compared to eggs, they are not free. Just as males who defend their genetic patrimony by mate-guarding are unable to simultaneously seek their own EPCs, those who employ frequent copulation may limit the amount of sperm they have available for gallivanting. Even the most super-stud males, after all, cannot produce unlimited amounts of sperm or semen. Because of their basic biology, they can be more profligate than females, but only within limits; to a degree, they, too, must be prudent. Hence, by giving themselves an advantage in sperm competition with their mates, such males might be placing themselves at a disadvantage when it comes to sperm competition for someone else's mate. Sure enough, when male rats, for example, are given the opportunity of mate-guarding, they deliver fewer sperm per ejaculation. So it may be that mate-guarding is doubly preferred, both to seeking EPCs and also to sperm competition.

Sometimes, however, males have little choice: Females mate with more than one male and cannot be prevented from doing so. What is such a male to do? In one particular species of zebra—known as Grevy's zebra, for its discoverer—individuals live in groups whose membership is constantly shifting. Females associated with a given male are likely to mate with a different male not long afterward. In fact, during a single day they may mate with an average of four different males. (Call these females polyandrous.) On the other hand, there are some Grevy's females—generally, those who have just given birth—who remain with one male for a prolonged period, during which they are essentially monogamous. They do this, by the way, because they need reliable sources of water, which are found only on a male's territory. So Grevy's stallions have two different kinds of females to deal with, those that are sexually faithful and those that aren't (bearing in mind that the same female will occupy different roles at different times in her life).

Grevy stallions adjust their tactics accordingly, depending on whether their female consort is polyandrous or monogamous. When mating with polyandrous females, males invest more time and energy in mating itself: Stallions call to and copulate seven times more frequently than when involved (temporarily) with monogamous females. They even ejaculate larger quantities of semen. It is also worth noting that in another zebra species, the plains zebra, females live in traditional harems, each led by single male, and as far as is known, they only mate with the harem-keeper. Plains zebra stallions copulate less, produce less semen, and also have smaller testes than their Grevy's counterparts, which have to be prepared to deal with females having an occasional penchant for a high-frequency of EPCs.

Grevy's zebras do not mate-guard; instead, the stallions are prepared to engage in sperm competition when need be. This raises new questions. Why rely on one strategy rather than another? Specifically, why don't all animals mate-guard, since that seems more efficient? (And also, what about human beings, who are not shrinking violets when it comes to mate-guarding but who also copulate far more frequently than is needed for reproduction alone?) The answer seems to be that in most cases mate-guarding is the primary strategy, with frequent copulation being the next-best alternative, resorted to when males and females must spend substantial time apart. This occurs, for example, among predatory birds; one individual often remains by the nest while the other goes on lengthy hunting excursions. Not surprisingly, predatory birds copulate a lot.

Take ospreys. These "fish-hawks" are large predators, the males of which provide virtually all the food while their mates are occupied with nest-site duties. Hence, male ospreys are unable to guard their females; they are too busy fishing. Female ospreys remain invisibly chained to the nest from the time they arrive in the spring until their young are independent, at about three months of age. Observations at several osprey nests reveal that pairs copulate frequently, on average 59 times per clutch, beginning when the female arrives on the territory. Males are absent about 30 to 50 percent of daylight hours, which provides opportunity for female EPCs as well as the motivation for paired males to insist on frequent copulations when they are back at home after a long day's fishing.

In the case of purple martins, we saw older males using EPCs to take reproductive advantage of younger, inexperienced males. In other species, males are likely to obtain EPCs from females who are poorly provisioned by their mates. This has been especially well established among our friends the ospreys. So, not only does a hardworking male osprey run the risk that his mate will "take a lover" while he is away looking for food, but that risk is intensified if he is no great shakes at bringing home the salmon. It is not known whether male ospreys that are particularly inept as providers try to make up for this by copulating even more often than is the osprey norm.

It is well known, however, that among many different species, a resident pair is especially likely to copulate just after an intrusion into the pair's territory. It appears—although it is not yet proven—that this response to intruders is initiated by the in-pair male. This would make biological sense, since intruding males in such cases are unlikely to have simply dropped by to borrow a cup of sugar or to sell Girl Scout cookies. Why should the female go along with this, agreeing to copulate with her mate just because some other male has recently been hanging around? Perhaps it is simply less costly for her to acquiesce than to resist her mate's importunities. Or it may pay her to permit copulations—especially when her fidelity is in question—so as to persuade her mate of his paternity, in order to assure his assistance in rearing the young. (As we shall see, the loss of paternal assistance is a major potential cost to females of EPCs, if discovered by the in-pair male.)

In any event, there are few things as sexually stimulating to socially monogamous animals as the possibility that the mated female might have had an EPC. Among orioles, males will copulate with their mates immediately after hearing a recorded song from another oriole. One might say that in the oriole world, the song of a male is sexually arousing to other males; the evolutionary significance of this would be that the nearby song of another male suggests that someone might have recently copulated with the in-pair female. If so, paired males who are "turned on" by this telltale signal and who introduce their sperm as quickly as possible to compete with the extra-pair male would be favored by natural selection over those paired males who were indifferent to such cues.

David's research on mallard ducks has shown that males (especially unmated bachelors) will often "rape"—that is, force a copulation upon—already-mated females. When this occurs, the mated male will typically respond by attempting to dislodge the attacker. Very quickly afterward, he forces a copulation with his own mate. Such behavior is not gentlemanly, but in the cold calculus of natural selection, it may be the best he can do to attempt to counteract the recent extra-pair mating.

The Galápagos hawk is unusual in that it is socially polyandrous: Up to five males will bond, socially and sexually, with one female. As soon as one male copulates with the female, the others quickly line up to do the same. It is not enough to say that sex, like laughing or yawning, is a "contagious" behavior. The best explanation for *why* it is contagious is that it signals the prospect of an extra-pair copulation and thus arouses an adaptive response, especially among males.

Mammals are not immune. Among rats, males mate with a female as quickly as possible after she has finished copulating with a prior male. Among nonhuman primates, males quickly mount and copulate with a female partner who has recently copulated with a different male. This implies that males are sexually aroused by indications that a female has recently copulated. And this, in turn, should not be very strange: Human beings, especially males, are also highly aroused by such indications. Hence, the attractions of hard-core pornography and voyeurism, which have been attributed to males generally having a low threshold for stimulation (since their investment in sperm is minimal).

This explanation is valid, as far as it goes, but it may not be complete. Thus, the adaptive significance of sperm competition may also be involved, since it would be adaptive for human males—no less than males of the other species we have mentioned—to be especially aroused by the prospect of sexual intercourse itself. This is even true if the sexual action is on the part of other individuals, if it indicates the nearby presence of a receptive female. Sperm competition would make it worthwhile for males to be prepared to join in, if possible, and to do so promptly, so as to compete with the preceding males. Moreover, the existence of sperm competition also helps explain a seemingly peculiar yet widespread aspect of human sexuality: Many men are sexually aroused by thinking of their female partner having sexual relations with another male. Some even go out of their way to arrange such encounters (although this appears to be rare).

In at least one rat species, when a male has recently copulated with a female, the time interval before his next copulation with that same female is significantly reduced if, in the interim, he observes her copulating with another male. Most likely, by copulating promptly with a female who has recently copulated with another male, the "responding" male increases the chances that he, instead of his rival, will fertilize the eggs. Alternatively, he may simply diminish the chances of his rival(s) being successful, if there is direct interference among sperm produced by different males.

And so, we come to an indelicate but revealing subject: testicle size. Species in which mate-guarding predominates generally have small testes; when frequent copulation is the preferred strategy, then, not surprisingly, the resulting male gonads are far more impressive.

For notable cases of frequent copulation, take those species that are polyandrous, in which one female regularly mates with more than one male (such as the jacanas, mentioned earlier). These animals typically have a high copulation frequency, probably initiated by males, each attempting to swamp the sperm of his rivals and thereby increase the chances that the eggs to be deposited in his nest—and which he will then incubate and care for—are genetically his. The males of such species have oversized testes, producing more sperm than monogamous males whose responsibility is to inseminate their mate but not to compete with the sperm of other males. This pattern is not limited to birds. It has been found for mammals generally, confirmed by comparing, for example, rodent species that do and do not have many EPCs, as well as members of the horse family (including zebras) and even balleen whales. Ditto for primates.

As already mentioned, some bird species, including prairie chickens and sage grouse in North America, breed on what is called a "lek," a communal displaying ground. Males gather here and show their wares, calling and posturing and typically arranging themselves in a dominance pattern, with the alpha males in the middle. Females mate almost exclusively with these favored individuals, who may copulate with numerous females in one day; these females, in turn, generally give the subordinate males a cold shoulder. Systems of this sort provide the opportunity to answer this question: When males have especially large testes, is it because of sperm competition (that is, because other males are likely to be copulating with the same females) or simply because of the demands of producing enough sperm to fertilize the eggs of many different females?

The answer is pretty clear: Whereas polyandrous species have large testes, the males of lekking species have testes that, corrected for overall body size, are if anything exceptionally *small*. Evidently, it is rather easy to make enough sperm to fertilize one female or even many, as dominant lekking males do. What really makes for big balls is when males must compete with the sperm produced by other males.

The most impressive case—or, at least, the one closest to home for readers of this book—comes from observations of the great apes. Remember those impressive silverback male gorillas, large in body and relatively aggressive in temperament, who succeed in dominating other males and gaining reproductive rights to a small harem of females? Although their bodies are large, their testicles are remarkably small, indeed downright tiny once corrected for body weight. By contrast, chimpanzee males—which do not achieve anything like the reproductive despotism enjoyed by their gorilla counterparts—have immense testicles. This is entirely reasonable, since a female chimp in heat will copulate with many different males; in one case, Jane Goodall observed a female chimpanzee copulate 84 times in eight days, with seven different males. As a result, a male chimp cannot simply assume that sexual access to an estrous female will result in paternity. He must produce enough sperm to give them—and thus, himself—a fighting chance. (When it comes to testicle size, human beings fall somewhere between the polygynous gorilla and the promiscuous chimp, suggesting that we are mildly polygynous. More—much more—on this later.)

Since testis size is largely influenced by sperm competition, not the simple need to inseminate a faithful partner, it is possible to use testis size as a rough measure of the sperm competition experienced by a species. In socially monogamous animals, there is much variability: Some species have relatively small testicles, suggesting very few EPCs. Others—especially those living in colonies—have large testicles: These include herons, sparrows, most seabirds, bank swallows and cliff swallows. In these cases, paired females have the opportunity to mate with other males. Of course, the same is true for males: They have the opportunity to mate with other females. But as we have just seen, it is unlikely that testis size is driven as much by the need to make lots of sperm in order to inseminate other females as by the need to compete with the sperm of other males. From this perspective, the female reproductive tract is an arena within which some pretty fierce sperm competition takes place. The tactics can be downright weird.

For example, it isn't even necessary to make large amounts of normal sperm, and in some cases males are better off if they don't, especially if sperm manufacture is relatively costly. Like an innkeeper trying to stretch his budget by watering down the drinks, males of some insects can induce a female to be sexually unreceptive to other males, at low cost to themselves, by introducing "cheap filler" into the females' seminal receptacles. Among certain species of *Drosophila*—fruit flies—males produce at least two different structural types of sperm, short ones and long ones. (The technical term is *sperm heteromorphism,* literally "different structures.") The assumption is that short sperm are cheaper to produce than long ones; this seems likely, since males producing short sperm mature earlier, suggesting that they have been using less metabolic resources. Although the shorties often constitute more than one-half the ejaculate, it appears that they do not directly fertilize eggs; their exact function is unknown, although the favored hypothesis is that the short sperm act as cheap filler within the female genital tract, making it less likely that a mated female will attempt to copulate again in the near future.

Among many insects, females are equipped with various organs that are specialized to receive sperm during copulation. Females will often remate when their sperm supply begins to wane. In some cases, it is simply the stretching effect, not some fancy pheromone, that does the trick, not unlike the hunger pangs that people feel when their stomach shrinks and the feeling of fullness and satiation that comes after eating, when the stomach is stretched. One researcher tried injecting silicone oil into the bursa copulatrix of a species of butterfly; in response, the females became not only distended, but sexually unreceptive. It is interesting to note that among butterflies, so-called apyrene sperm (those lacking genetic material altogether) are especially common, sometimes exceeding 90 percent of all sperm produced. It seems likely—although as yet unproven—that it is metabolically cheaper for males to produce these "blanks" than to outfit all their sperm with the full complement of DNA.

In the annals of sperm competition, an important consideration is whether there is an advantage in being the first to mate with a given female, or the last to mate, or whether all the sperm accumulate within a multiple-mating female, resulting in a "random lottery" with likely victory going to the male who contributes the most. Among insects and birds, the predominant pattern is "last male advantage," or "last in, first out." When sperm are accumulated inside the female in a specialized sperm storage organ (as in *Drosophila* or butterflies), it makes sense that the last sperm to have been added are likely to be the first squeezed out when her eggs finally pass by in reproductive procession. Like airline passengers waiting at a baggage carousel, whose luggage comes out sooner if it went in later, the last are likely to be first. When it comes to the reproductive trafficking of some males, this has important implications for behavior.

Imagine that you are a male in one of these "last in, first out" species. Add to this the fact that ecological circumstances may dictate that you and your mate must spend considerable time apart. Not knowing for certain whether she has copulated with anyone else in the interim, your best strategy is to copulate often. That way, you increase the chances that yours will be the most recently deposited sperm—hence, the ones most likely to fertilize your mate's next egg.

All this assumes, of course, that the male and female have already made a reproductive commitment to each other—through courtship, building a nest, perhaps jointly defending a territory, and so on. In other cases, when EPCs loom large in evolution's strategic planning but the prospective pair have not yet decided to settle down together, there are alternatives to mate-guarding or copulating often. One of these is to refrain from making a commitment. For example, male ring doves behave aggressively toward females that reveal by their behavior that they have recently copulated with other males. The result is to delay the pairing, which is probably adaptive, since it means that by the time a male ring dove commits himself to a female with a colorful past, she will have already revealed that past by laying fertilized eggs. It is interesting that ring doves have an unusually short duration of sperm storage, so the suspicious male does not have long to wait.

What is a penis for? Ask a young boy and you get an unequivocal answer: peeing. (Girls, who manage quite well without one, have good reason to disagree.) So, what is a penis really for? Ask an adult and you'll probably be told that it is for introducing male reproductive products into the female. Once upon a time, long, long ago—when the myth of monogamy still reigned—nearly all biologists would have agreed. But no longer.

Don't misunderstand: The penis *is* for introducing male reproductive products into the female, but that isn't all that it does. Since we now understand that females of many, probably most, species are likely to mate with more than one male, the corresponding likelihood is that these males are not simply adapted to transferring sperm . . . even lots of sperm. If you are the second, third, or fourth male to mate with a female, your long-term reproductive interest would be served not only by being able to introduce sperm efficiently and in adequate amounts, but also by being able to remove any rival's previous deposit. In many animals (especially insects), the penis is not merely a pipeline for delivering sperm; it is also variously a scraper, gouger, reamer, corkscrew—a veritable Swiss Army Knife of gadgets and gizmos adapted to removing the sperm of any preceding male.

In the black-winged damselfly, a common streamside insect of the eastern United States, females commonly mate with more than one male. Each male black-winged damselfly sports a specialized penis outfitted with lateral horns and spines, not unlike a scrub brush. Copulating males use their penis to clean out from 90 to 100 percent of their predecessors' sperm before depositing their own. Some male sharks give their sexual partners a kind of precoital douche, courtesy of a remarkable double-barreled penis. One barrel contains a specialized tube that can act as a high-pressure saltwater hose, sluicing away any sperm deposited by a sexual rival; the other barrel transports sperm into the female. Male pygmy octopuses apparently can detect whether a female with whom they are mating has already copulated, because they spend more time coupling with already-mated lady octopuses: They spend the extra time using a specialized tentacle to scoop out sperm deposited by their predecessors.

We have already described how in most insects, females store transferred sperm in a special organ, a spermatic receptacle, bursa copulatrix, or spermatheca. Then, as an egg passes along its own conduit, the female contracts special spermathecal musculature, forcing out sperm that fertilize it. Not surprisingly, males of some insect species take advantage of this arrangement: Rather than removing a rival's sperm directly, they induce the female to do it. Prior to mating, they employ specialized genital structures to stimulate the female's sperm-ejection system, so that their spermathecae contract, ejecting previously stored sperm but in the absence of any eggs.

Don't miss the forest for the trees: This is not simply a recitation of Barash's "Believe It or Not" compendium of sexual oddities. These examples all reflect the powerful action of evolution, conveying an advantage to males who are capable of overcoming their rivals' sperm. Males have stumbled onto other techniques for giving their sperm an advantage in the reproductive fray. For example, a small European bird called the dunnock mates occasionally in pairs, and sometimes in threesomes, of both sorts: two males and one female, and two females and one male. (For a drab-looking bird, the unprepossessing dunnock is a bit of a swinger.) When it's two dunnock males and one female, the males typically peck the female's cloaca before copulating with her; in response, she squeezes out a few drops of her other husband's sperm. The more time a female dunnock has spent with one male, the greater the number of precopulatory cloacal pecks delivered by the other one.

Finally, there is traumatic insemination. The best-known examples are found among bedbugs; the males simply pierce the body of their mate/victim, injecting sperm that then travel through the blood, collecting in the gonads and achieving fertilization. But this is "merely" a way of inseminating females, not sperm competition per se (unless the sperm of two of more males battle it out within a female's bloodstream, something that has not—yet—been demonstrated). There is, however, at least one bizarre example of males using traumatic insemination against each other; it involves a male hijacking another's reproductive efforts, achieving sperm competition with a vengeance. There exist bugs that live on bats that inhabit caves. Among these aptly named cave bat bugs, males attack other males, injecting sperm as well as seminal fluid directly into the victim's body cavity, which is pierced by the attacker's sharp penis. The male recipients metabolize the seminal fluid, thereby gaining some calories from the transaction. But some surviving sperm also migrate to the recipient's testes. If and when the victim copulates with a female cave bat bug, he will therefore transfer some of the sperm of his attacker, who gets paternity by proxy.

By this point, it should be clear that males work awfully hard to obtain EPCs and, similarly, to prevent rival males from getting any at their expense. For animals, at least, the underlying reason for all this mate-guarding, gallivanting, and sperm competition is the proverbial bottom line: not financial profit, but an evolutionary bonus in the form of reproductive success. A male's reproductive success is gravely threatened if his mate has an EPC or—worse yet—an EPF (extra-pair fertilization). To put it bluntly, there is no payoff to rearing someone else's offspring.

What options are open to a male who has been cuckolded? Not very many, and all carry substantial liabilities. They include:

1. Physical aggression: Punish your mate and/or the interloping male. This makes a certain intuitive sense but is unlikely to convey much evolutionary benefit, except possibly if it makes either party less likely to transgress again in the future. It also risks physical injury as well as disrupting the current mateship.

2. Force a prompt copulation with your mate, in the interest of sperm competition. This, too, might weaken the pair-bond.

3. Provide less paternal care, possibly even neglecting or abusing any succeeding offspring. This might backfire as well, especially if some of those disadvantaged youngsters include your own.

4. Desert your mate altogether. This only makes sense if alternative mates are available. Not surprisingly, adultery is frequently cited among human beings as the reason for a divorce. It may surprise some readers—although at this point, perhaps not very many!—that divorce also occurs among animals and that here, too, it is closely associated with EPCs, especially on the part of the female.

Among the above options, number 3 seems especially prevalent: When males have indications that their mates have been unfaithful, they seem particularly unlikely to act as devoted fathers. We expect that males who follow a "hit-and-run" strategy, contributing sperm but little else to their offspring, would be considerably less troubled if their mates engage in EPCs. Or, at least, since they are not otherwise inclined to be devoted fathers, departures from monogamy on the part of their sexual partners are unlikely to have much effect on the males' behavior. But ostensible monogamy is a different matter. In these cases, males typically give a lot— parental care, plus typically forgoing large numbers of their own EPCs— and, not surprisingly, they expect to get a lot, or, at minimum, to "get" to rear their own progeny.

True to expectation, male barn swallows are less attentive in feeding chicks if their mates had previously engaged in EPCs. Among indigo buntings, yearling males are approximately twice as likely as older males to be cuckolded by their mates; significantly, only rarely do these younger males help out in rearing chicks. These findings make perfect biological sense. After all, it is unheard of, and indeed almost unimaginable, for adults to dispense parental care randomly with regard to genetic relationship. Try to imagine a society—animal or human—in which perfect strangers care for the young, essentially at random. Such an arrangement has never been found for any living things. To be sure, there are many species in which males contribute little or no paternal care, but when they do act paternally, then they, no less than females, direct their attention to particular mates, particular nests, and particular offspring: their own. The next step is to withhold some or all of that attention when there is a good chance that the offspring in question are someone else's. (This doesn't require any special intellectual insight on their part; any genetically mediated tendency to provide care indiscriminately would quickly be replaced by alternative tendencies to direct care to one's own offspring . . . which, because they are one's offspring, are likely to possess the genes for such tendencies, which would therefore be given a boost.)

As already noted, monogamy—even social monogamy—is rare among mammals. It is noteworthy, however, that it is essentially only in cases of monogamy that male mammals provide *any* paternal care; after all, even with the danger of EPCs, a monogamous male mammal has much more confidence of his paternity than one whose sexual partner is likely to have been inseminated by one or more other males. And so we find comparatively devoted fathers among foxes, coyotes, beavers, gibbons, and marmosets, species in which the females are by and large sexually faithful to just one male. And we find very little paternal behavior in woodchucks, porcupines, squirrels, deer, wildebeests, cougars, or bears. In proportion as father-hood is in doubt, paternal behavior is likely to be lacking.

At other times, the prospect of parenthood seems capable of generating parent-like behavior, masquerading for a time as disinterested altruism. The emergence of EPCs as a major fact of life has even diluted the impact of another revolution in modern biological theory, known as "kin selection." Numerous perplexing cases have been recorded in which young adults serve as "helpers at the nest," assisting others to reproduce rather than rearing their own offspring. With our recent appreciation of EPCs, a new wrin-kle has emerged in interpreting such seemingly altruistic behavior. Even "helper" males may actually be helping themselves . . . to occasional sex with the breeding female. So-called cooperative breeding may therefore involve less altruism than had recently been thought, since, in at least some cases, what looks like altruistic baby-sitting is actually full-fledged parental investment, provided to offspring of the helpers themselves, some of whom were conceived via circumspect EPCs.

And yet, confidence of relatedness doesn't explain all aspects of parent-ing. For example, an experiment looked at the paternal behavior of male common gobies; these are small marine fish, the males of which guard eggs left them by a female. The goal of the experiment was to assess whether male common gobies treat their offspring differently depending on whether they had spawned alone (with a female but no other male) or with a second male present, in which case there was at least a chance that some of the eggs and fry were fathered by the interloper. It turned out that it didn't matter whether a second male had been present.

The gobies are not alone. It is not invariant that paternal care varies with confidence of paternity; there are some notable exceptions, not only among fish but among birds as well. What are we to make of these exceptions? (Make no mistake: Exceptions they are.) If they were the rule, then we'd have to reassess some basic evolutionary and genetic principles. As excep-tions, they provide us with the opportunity to fine-tune our predictions.

In the case of occasional paternal care by nonpaternal animals, other fac-tors appear to be at work. For example, if EPCs don't usually occur in a given species, then, lacking the context, there would be little or no evolu-tionary pressure selecting for a male's ability to detect his likely nonpater-nity and to react accordingly. Natural selection can only generate a response if, in the past, situations have arisen that cause individuals responding in one way to be more successful than individuals responding in other ways. By the same token, we lack the ability to hear ultra-high-frequency sounds because such sounds have not been part of the relevant landscape in which our ancestors evolved; the same is not true of certain moths, however, which have evolved the ability to hear and respond to the ultra-high-frequency sounds emitted by a highly relevant part of their environment: hunting bats.

It is also possible that, in some species, males simply lack the ability to detect eggs or offspring not their own, even though it might be to their benefit if they could. Or maybe in certain cases the payoff that comes with discriminating "my genes" from "someone else's" is substantially reduced by certain disadvantages, such as the costs of occasionally erring and discriminating against one's own offspring after all. Nonetheless, there is little doubt that incursions—especially by a male into the territory of a monogamous breeding couple—are not appreciated . . . especially by the resident male.

In the mid-1970s, David conducted this experiment in Mount Rainier National Park: He attached a model of a male mountain bluebird near a female and her nest, so that when the female's actual mate returned, he discovered his female in close association with this apparent stranger. The male behaved aggressively toward the model and also toward his own mate, in one case driving her away; she was eventually replaced with another female, with whom he successfully reared a brood. This little study became somewhat controversial, with researchers debating, among other things, the propriety of biologists acting as Iago and inducing violent sexual jealousy on the part of their subjects! In any event, there have since been numerous studies in which males of different species were removed while their Desdemonas were sexually receptive, in which these males had visual access to their unguarded females, in which the females were removed, in which males witnessed their females in cages with decoy males, and so forth, all looking for possible impact on the males' subsequent behavior, especially his paternal inclinations.

The pattern persists: Genetic paternity correlates with acting paternally. But not always. Especially among some socially monogamous species, males do not consistently reduce their paternal solicitude following behavioral evidence of their mates' infidelity. Maybe they just don't "understand" what has happened, or perhaps their paternal inclinations are so hard wired that they simply don't have enough flexibility to adjust. In any event, it is interesting that the strongest evidence for precise adjustments by males to the EPCs of females comes from cooperatively breeding species such as the dunnock, where several males might be associated with one female. Here, males care for offspring in proportion to their likelihood of being the father; if several males have copulated with one female, each male will provide food, for example, proportional to his degree of sexual access. (More copulating, more food-bringing.) In another cooperatively breeding bird species, the acorn woodpecker, when dominant males are experimentally removed from the group, they respond by infanticide, destroying eggs laid while they were out of the reproductive picture. The likelihood is that in such species, males are often exposed to variations in the probability of being fathers; hence, they have the behavioral repertoire to detect such probabilities and to behave accordingly. Perhaps in cases of ostensible monogamy, females are normally so adroit at hiding their EPCs that males have not evolved a response.

There are other interesting avenues connecting EPCs and parental behavior. We have already looked at the peculiar trade-off between mate-guarding and gallivanting, with one precluding the other. Males also appear to be influenced by another balance point: between gallivanting (going in search of EPCs) and staying home to help take care of the kids. As with gal-livanting versus mate-guarding, males can't have it both ways; if they are off trying to spread their seed, they cannot very well also be home tending the fruits of that seed.

In many colonially nesting bird species (e.g., terns, herons, social gulls), there is comparatively little EPC activity, perhaps because the females breed synchronously; that is, they are all likely to breed at about the same time. As a result, a male who gallivants runs the risk that his own female will cuck-old him. By contrast, most songbirds appear to engage in EPCs if they can. Although they breed seasonally, they are not truly synchronous, so a male can inseminate "his" female, guard her against other males while she is fer-tile, and then proceed to seek other females who may be just entering their fertile period. Also, many males seem to use a "switching" strategy: After their eggs have hatched, they abandon gallivanting and become doting fathers . . . because at this point the genetic payoff exceeds that from seek-ing EPCs.

For example, male indigo buntings seek EPCs while their mates are incubating—at a time when there is relatively little that the males can do to aid their offspring. Paternal behavior competes with trying to get EPCs: In most species of birds, males provide quite a bit of parental care during chick-rearing, much less during nest-building or incubation. It may be no coincidence that in these early stages of the breeding cycle, males have the prospect of achieving one or more EPCs; hence, they are more likely to gal-livant. By the chick-rearing stage, most fertile females have already been inseminated, so the best thing a male can do is help rear the offspring he has (presumably) fathered.

On the other hand, more effort is probably required to rear hungry, fast-growing chicks than to build a nest or sit on the eggs. So male birds may put more effort into chick-rearing simply because, at this point, females are less able to succeed as single parents.

The decision for many males comes down to this: Seek EPCs or be a stay-at-home parent. It is a trade-off between two kinds of striving: mating effort (trying to obtain as many copulations as possible) versus parental effort (trying to enhance the success of those copulations already achieved). Males typically do whichever offers a better return. For example, if there are fertile females nearby, EPCs—mating effort—may be favored; if there are lots of predators, parental effort; if there are lots of other gallivanters, mate-guarding combined perhaps with parental effort; if your offspring have especially high metabolic needs, parental effort; if your mate has likely copulated with other males, less parental effort and more mating effort (with that same female or others); and so forth.

Earlier, we encountered hoary marmots, caught in the dilemma of whether to mate-guard or gallivant in search of EPCs. Sometimes these animals live rather isolated lives, the basic unit consisting of one adult male, one or two adult females, and their offspring, with no one else nearby. In other situations, hoary marmots occupy bustling colonies, such that although a male is likely to mate with one or two nearby females, there are also many additional females—and males—in the immediate vicinity. It turns out that isolated males are rather good fathers, highly attentive to their young, whereas those occupying busy colonies spend their time wandering about in search of EPCs or mate-guarding, defending their females from other males in search of EPCs. Their offspring get short shrift.

We have long known that there is considerable variation in the extent of male parental care; generally, those species more inclined to monogamy are more likely to be good fathers. Recently, it has become clear that there is also quite a range of paternal behavior within most species as well. Attractive males usually provide less parental care, so that females end up doing relatively more mothering when they are paired with "hunks." This tendency is captured in the seemingly dry title of this scientific article: "Paternal Contribution to Offspring Condition Is Predicted by Size of Male Secondary Sexual Characteristic." The greater the male's secondary sex characteristics, the less his contribution. It is as though desirable males know they are desirable, and so they are likely to shop that desirability around; by the same token, those "lucky" females who get to mate with such studs find themselves less lucky when they are stuck with most of the household chores.

Imagine, for example, a type of bird in which males with bright red spots are especially successful in seducing females. Now imagine a male whose spots are particularly bright and red: Because he is so sexy, his efforts at EPCs are likely to bear fruit, and so he spends most of his time gallivanting about, leaving his mate to pitch in with the kids to make up for the deficit.

Don't expend too much pity on Mrs. Stud, however: First of all, it was her choice to mate with the lazy, conceited jerk, and second, in all likelihood she will profit genetically from the transaction, since her sons will probably inherit their father's dashing good looks—as well as his lousy paternal habits—and also, therefore, his attractiveness to a new generation of eager females. As a result, the hardworking female will likely have more grand-children via her male offspring.

Incidentally, long-tailed male barn swallows fly less efficiently than their short-tailed counterparts, so the fact that such males are more inclined to be dead-beat dads may be due at least in part to the fact that it is more difficult for them to perform the normal activities of barn swallow parent-ing—specifically, catching insects on the wing and bringing them back to their brood.

By contrast, comparatively unattractive males are more inclined to be good fathers. It appears that they make the best of their bad situation by behaving as paternally as they can, even though some of the offspring thus aided are not their own. Among purple martins, for example, young males—especially prone to being cuckolded by their mates and also unable to recoup much by way of obtaining their own EPCs—behave as paternally as do older males, despite the fact that their payoff is lower (since many of their nestlings will have been fathered by those older males). They simply have nothing better to do.

Females don't always share the male enthusiasm for EPCs. And so, we come to rape. Some biologists prefer the more genteel phrase "forced copulation." But anyone who sees the phenomenon among animals is unlikely to have much doubt what is going on. Normally, sex among animals involves an extensive sequence of courtship interactions that are clearly consensual, in which the two participants bow, nod, sing, burble, prance, twist their bodies, arch their backs, curve their necks, exchange ritual food items, do a nifty little dance, clatter their beaks, bills, or muzzles together, wave their arms or wings, flap their ears or flare their nostrils, prance and strut, preen and groom each other, bill and coo in romantic synchrony, and, all in all, make music together, which, even when not beautiful, is at least mutual. In short, they go through a rather elaborately choreographed and predictable pattern that eventually results in their becoming sexual partners. If one would-be lover misses a cue, or otherwise behaves inappropriately, the courtship may be broken off. Assum-ing all goes well, however, the mated pair eventually copulates, and al-though their coupling may not always meet the human definition of "romantic," it is at least likely to be well synchronized, smoothly accomplished, mutually arrived at and agreed to—a pattern of consent that is further underlined if the partners remain together in some kind of social bond (such as we identify with monogamy).

By contrast, it is very different thing when one or more males descend upon a female, whether mated or not, and immediately force a copulation, which generally includes ejaculation, without a "by your leave." The female typically struggles vigorously and may sometimes escape; her mate, if present, generally tries to drive away the attackers. No subsequent social relationship is established between the violators and the victim, who is not uncommonly injured in the sexual attack. Sometimes fertilization results. If this isn't rape, what is?

I and others have documented a violent and brutal pattern of forced copulation among mallard ducks, for example. The act occurs most commonly when the drake is some distance away, and it unfolds much like a gang rape among human beings. A small flock of males swoops down upon a hapless female; the victim struggles vigorously, trying to escape. Neither she nor her attackers engage in any of the shared niceties that characterize typical courtship between a pair of mated mallards. And no subsequent social relationship is established. Although females thus attacked are sometimes drowned in the process, raped mallards often survive and bear their victimizers' offspring.

Behavior of this sort has been described extensively among animals as diverse as fruit flies, mole crabs, scorpionflies, crickets, desert pupfish, guppies, blueheaded wrasse, bank swallows, snow geese, lesser scaup and green-winged teal (duck species), African bee-eaters, laughing gulls, tree shrews, elephant seals, right whales, bighorn sheep, and feral dogs. This list will undoubtedly grow as the number of long-term behavioral studies on animals increases. To some extent, it is also a matter of taking a clear-eyed look at phenomena that have already been well "known" (that is, described) but not correctly interpreted. For example, it has long been known that up to a dozen or so male house sparrows will often congregate around a female. These agitated gatherings of males used to be called "communal displays" before their true nature was recognized. They are multi-male EPC attempts. Since in nearly all cases the female resists, they could also be called "gang rapes."

In any event, there is simply no question that Susan Brownmiller, author of the best-selling book *Against Our Will*, was flat-out wrong when she asserted that rape is unique to human beings. It is not.

Rape also appears to be common among primates, having been reported in rhesus monkeys, talapoin monkeys, vervet monkeys, stumptail macaque monkeys, Japanese macaques, spider monkeys, gray langurs, gorillas, chim-panzees, and orangutans.

In a sense, such cases are fair game for this book, since they represent departures from monogamy. When female ducks are raped, this might seem yet another type of EPC; after all, they are indeed extra-pair copulations and can result in mixed paternity. The threat of rape may even motivate a substantial proportion of animal mate-guarding, just as its reproduc-tive consequences may well include a reduction in the mated males' incli-nation to behave paternally toward any offspring conceived as a result. But for our purposes, it is more interesting to focus on those situations in which those individuals going outside of monogamy are doing so "of their own free will."

This is a difficult and slippery slope, however, for several reasons. First, free will among animals (not to mention people!) is a much-contested topic. Scientists generally steer clear of it, preferring to deal with what animals actually *do,* instead of whether or not they have any choice in the matter. Second, although being a rape victim seems an extreme case of being deprived of one's autonomy, an argument can be made that all the decisions an animal—or a person, for that matter—makes are done under duress: If a male songbird goes in search of EPCs "because" he carries a genetic ten-dency to behave this way (because his father did, which earlier contributed to his being conceived), is he really acting of his own free will? And if a female blackbird copulates with a neighboring male in order to gain access to his food supply, or a female barn swallow does the same in order to gain access to a particular male's genes, isn't she also the victim of a kind of coer-cion? Nonetheless, there is a common-sense distinction to be made between coercion orchestrated by the conflicting will of another individual (rape, social subordination, etc.) and coercion that results from the pressure of cir-cumstance (e.g., shortage of suitable food or genes).

Just as there are sometimes disputes about whether sex was consensual or forced among human beings, there are gray areas in certain animal cases, too. Males are sometimes aggressive in seeking EPCs, verging on rape, as in the case of indigo buntings. In others—such as black-capped chickadees and blue tits—females actively solicit copulations, while in yet others, males are unquestionably not only initiators, but brutal assailants.

In any event, unattractive and otherwise unsuccessful males are in an especially difficult situation. We have noted that compared to females, the parental investment of males is relatively skimpy, so they are expected to compete with other males and/or to be attractive to females. But what if they are uncompetitive and unattractive? Even if they succeed in pairing with a female, as we shall see in the next chapter, their mates will often take the opportunity of improving their reproductive situation by mating on the sly with more desirable males, whereas these unappealing males may have no comparable recourse, except for rape, since they are likely to be rejected by females if they attempt EPCs, for the same reason that they are liable to be cuckolded by their own mates. There is in fact a growing body of evidence that human rape, too, tends to be a reproductive tactic of likely "losers."

We have already seen that when rape occurs among mallards, paired males often attempt to defend their mates, and when unsuccessful, they frequently respond by raping the victim themselves. This has subsequently been found for many other species as well. Why such an ungentlemanly response? Probably because, given "last male advantage" as well as the possible payoff of simply diluting any sperm introduced by the rapists, the mated male is attempting (albeit unconsciously) to increase the chance that he will father any chicks produced.

It is probably no coincidence that mountain bluebird males attacked their mates after they had been deceived as to their fidelity; among this species, there is typically a reservoir of available, unmated females. By contrast, mallard drakes whose females had been raped nearly always remain mated to them; in their case, there is generally a shortage of unmated females, so a drake is better off remaining with a sexually compromised mate than abandoning her and likely ending up with no mate at all.

Actually, the highest recorded frequencies of rape are found in a closely related pair of geese, Ross's goose and the lesser snow goose. Research conducted at the largest known goose colony in the world—consisting of 291,000 Ross's and 297,000 lesser snow geese, on Karrak Lake, Nunavut (formerly part of the Northwest Territories, Canada)—has found that among these attractive, innocent-looking birds, about 50 percent of attempted copulations occur outside the pair-bond, nearly all of them the consequence of rape. About one-third of these rape attempts are successful, in that the attacking male achieves cloacal contact with his victim. But "success" in a biological sense is more elusive: Extra-pair paternity in the two species is consistently less than 5 percent, mostly because rapes occur too late in the breeding cycle, when the victimized females are no longer fertile. Evidently, rape is not an efficient reproductive tactic for male geese, although it pays off in some cases. Even though only 1 attack in 20 produces offspring, it can still be a winning strategy for the rapist if the costs to him are sufficiently low.

Incidentally, not all waterfowl experience high levels of rape. It appears that species that defend a relatively large territory around their nests are less likely to have to deal with sexually motivated intruders. Maybe this is one reason for defending a large territory in the first place. Good fences—

or, rather, widely separated territories—may make sexually well-behaved neighbors.

The upshot of all this: Even though monogamy is the primary mating system in mallards, geese, and many of the so-called puddle ducks, the male inclination for additional sexual opportunities has—to coin a phrase—muddied the waters. Just as it has for pretty much all other living things.

Soon, we'll look at our own species. But let's end this chapter by noting that human beings show the entire repertoire of EPC-related male behavior: a penchant for extra-pair copulations, a tendency for mate-guarding, frequent copulation, reduced paternal care in cases of reduced confidence of genetic fatherhood, maybe even anatomic adaptations of the penis and EPC-sensitive adjustments in sperm production. People are also prone to respond to real or suspected adultery by their mates with desertion and even violence. Indeed, adultery—or the suspicion of adultery—is a major cause of divorce, and of spousal violence as well. About one-third of spousal killings in the United States are due to female infidelity, whether correctly attributed or suspected. The frequency of infidelity-generated violence is, if anything, even higher in other societies for which data have been collected, such as Africa.

And yet, in spite of the high risks, female infidelity—along with male philandering—seems to be virtually universal. And not simply as a result of male compulsion.

The usual assumption among evolutionary biologists is that male sexual behavior is geared toward quantity of offspring, whereas its female counterpart is geared toward quality. This, in turn, is achieved by men being oriented toward *quantity* of sexual partners, and women, toward their *quality*. Monogamy, when adhered to, enforces a similar strategy on both men and women. It is in the realm of EPCs, on the other hand, that such male–female differences between quantity and quality are most likely to be revealed.

No one is claiming that males are always or single-mindedly in pursuit of extra-pair sexual opportunities—only that they are predisposed to do so under certain conditions. Furthermore, as we shall see, this seems to apply to *Homo sapiens* no less than to other species. On the other hand, traditional wisdom in evolutionary biology has claimed that females are comparatively coy, choosy, and faithful. They are. But increasingly, we are also learning that females in general—and this includes women in particular—are not so easily pigeonholed.

The myth of monogamy is seriously threatened, although monogamy as a human institution seems likely to carry on indefinitely, an ancient yet sturdy Potemkin Village behind its long-standing facade of polite fiction. There is little doubt that the majority of males, whether "married" or not, are favorably inclined toward out-of-pair sex. (This does not necessarily mean, however, that they will always act upon it; see Chapter 7.) For a "married" male to engage in out-of-pair heterosexual sex, his EPC partner must in turn be (1) seduced, (2) coerced, (3) a willing co-participant, or (4) an active initiator. And so, we turn to the role of the female. We'll find that all four patterns occur, in animals as well as human beings.

# CHAPTER THREE

# Undermining the Myth: Females (Choosing Male Genes)

S everal decades ago, a research team was looking into the prospect of using surgical birth control to reduce populations of unwanted birds. They experimentally vasectomized a number of territorial male black-birds and were more than a little surprised by the results: A large percent-age of the females mated to these sterilized males nonetheless produced off-spring! Clearly, there was some hanky-panky going on in the blackbird world: These females must have been copulating with other males, not just with their social mates.

Long before DNA analysis and the formal identification of EPCs, tanta-lizing findings such as this suggested that the traditional teaching among evolutionary biologists needed some revision. It had long been thought that females of most species were the "flip-side" of males: Their yearning for cozy monogamous domesticity was supposed to be about as strong as the male tendency to mate with as many different partners as possible. Whereas males were known to gallivant and try to sow their wild oats, their "wives," it was assumed, stayed home—at the nest or den—minding the hearth, duti-fully bearing young fertilized by their "husbands." The males had a fond-ness for philandering; females supposedly did not.

This expectation of a double standard in the animal world may have been soothing to the ego and also perhaps to the unspoken anxieties of many biologists . . . the majority of whom have long been male. But DNA fingerprinting and associated technologies have changed all that forever, confirming that, at least in some cases, females practice less than perfect sexual fidelity.

In *The Descent of Man and Selection in Relation to Sex* (1871), Charles Darwin wrote that "the males are almost always the wooers" and that "the female, though comparatively passive, generally exerts some choice and accepts one male in preference to the others." As we now understand it, Darwin was correct . . . as usual. But he must be taken more literally than one might think. Thus, it is not true that females accept one male and only one male, period. Rather, as Darwin pointed out, females accept one male in preference to the others . . . while often trying out the others, too! Female "preference," in this context, may mean giving an edge to the genetic contribution of one male rather than another, but this assuredly does not require monastic sexual fidelity on the part of the female. Coyness may have its value as public policy—the stance most females assume in front of strangers and, notably, their acknowledged social and sexual partner—but it does not necessarily reflect what they do in private.

Of course, the mere fact of extra-pair copulations does not in itself indicate female *choice:* In some cases, female animals no less than human beings are raped. But in many others, they actively solicit sexual relations with males who are not their acknowledged social partner. Female primates, for example, may temporarily leave their troop or—if ostensibly monogamous—their male "significant other" to hang out with one or more neighboring males; similarly, birds may fly onto the territories of other, already mated males, typically early in the morning. In such cases—and especially when genetic testing subsequently reveals that one or more of her offspring have been fathered by a male other than her presumed mate—there can be no mistaking either her behavior or her motivation . . . not unlike a woman who "just happens" to visit a man's hotel room late at night.

It remains true that the sexual tactics of males differ from those of females, being more showy, pushy, outwardly competitive, and sometimes even violent. In addition, just as not all males are philanderers, there are some cases in which females vigorously rebuff the extracurricular mating efforts of other males, suggesting that EPCs are not always in a female's interest. Nonetheless, the evidence has been accumulating, fast and furious, that females are not nearly as reliably monogamous as had been thought— and that often they are active sexual adventurers in their own right.

Why?

One possibility is that extra-pair copulations by females are nonadaptive, an unavoidable by-product of strong selection for multiple mating by males. Thus, perhaps any female penchant for EPCs is merely the equivalent of nipples in male mammals, a tag-along trait that has no value in itself but is maintained simply because it is advantageous in the opposite sex and somehow cannot avoid being expressed in both sexes, even though it is only meaningful in one.

An interesting idea, this, but one that is not supported by the evidence, especially since male and female mating tendencies are not genetically correlated; in other words, selection for high mating frequency in one sex does not necessarily produce high mating frequency in the other.

We are stuck with the question: Why aren't females more monogamous?

It is relatively easy to understand the evolutionary payoff that males derive from playing fast and loose. All other things being equal, more copulations mean more opportunities for them to project their genes into the future, and—usually—at relatively little cost. But what do females get out of EPCs? A tempting answer is simply that "they like it," or maybe "they find it exciting" or "interesting," or perhaps "it feels good." In evolutionary terms, however, these are all inadequate explanations, just as it is insufficient to explain sleep, for example, by saying that it is a response to being tired.

For sleep researchers, a crucial question is: Why does prolonged wakefulness make people tired? Tiredness is simply an internal state that leads people, under certain circumstances, to seek sleep; it is not an explanation in itself. Tiredness says nothing about the adaptive significance of sleep or about the mechanisms of sleep and, hence, why tiredness is a prelude to it. Similarly, for evolutionary biologists, it is not sufficient to say that females seek extra-pair copulations because they "like" it or because they sometimes find other males attractive. The crucial question is: Why do females of many species that are socially monogamous engage, at least on occasion, in extra-pair copulations? Another way of saying this: Why do they "like" it?

The one-size-fits-all adaptive explanation is that EPCs must somehow contribute to the ultimate reproductive success of the females that do it. The challenge, then, is to identify how this increased success comes about. What is the payoff to such females?

Broadly speaking, there are two kinds of benefits females can gain by mating with someone other than their identified partner. They can profit indirectly, via superior offspring fathered because of the EPC, or they can profit directly, gaining material benefits for themselves as well as their descendants. In this chapter, we'll examine the genetic or indirect benefits; in the next, we turn elsewhere.

One possibility—the simplest—is fertility insurance. To be sure, males produce millions, often hundreds of millions, of sperm in a single ejaculation, whereas, in most species, females release no more than a handful of eggs at a time. Just as the Marines used to advertise that they were looking for "a few good men," it seems likely that females, looking for a few good sperm, should have very little difficulty, since sperm are available to fertilize eggs in a ratio of millions to one.

Granted, it only takes one "good" sperm to penetrate an egg. Yet for reasons that are not clearly understood, it appears that there must be millions of willing and able pollywogs nearby in order for fertilization to be reliable. Among human beings, for example, if a man produces "only" 50 million sperm per ejaculation, he is generally considered sterile. In addition to sheer numbers, sperm must also be "viable," which means able to swim upstream—and quickly. Moreover, when they arrive at the Holy Grail of their journey—a ripe and willing egg—sperm have to be sporting the right kind of protein coat surrounding compatible genes. The upshot is that females who copulate with several different partners are more likely to get all their eggs fertilized; this has been found to be true for fruit flies as well as birds.

Female cormorants are more likely to engage in EPCs when they have produced relatively small broods. It is possible, of course, that EPC-prone females have fewer offspring *because* they have indulged in EPCs, but more likely that they have EPCs because they would otherwise have fewer offspring. This is suggested by the fact that females who failed to fledge chicks were those that subsequently engaged in most of the EPCs. Further evidence comes from a study of house sparrow nests, which found that nests containing infertile eggs are significantly more likely to contain an EPC-fathered chick than are those lacking infertile eggs. Females consistently have more EPCs when their breeding success is otherwise low, suggesting that it is not simply a matter of some females being particularly inclined to mate outside their social union.

Among those birds whose females are genetically rewarded for multiple matings are, interestingly, red-winged blackbirds, the same species that had so startled the ornithological world more than 25 years ago when females mated to sterilized males were found to have reproduced. Although the effect is not dramatic, it is nonetheless real: Female red-wings who "play around" lay more eggs and enjoy a higher hatching rate than those who remain sexually faithful to one mate. Moreover, their peak of EPCs occurs one day *closer* to egg-laying—when their fertility is higher—than the peak of IPCs (intra-pair copulations). Since there is no evidence that such females are being raped or coerced in other ways, the conclusion seems unavoidable: Female red-wings are timing their EPCs to maximize fertilizations (we'll see later that there is some evidence that human beings—probably unknown to themselves—do the same thing).

Although, under natural conditions, very few red-wing males are permanently sterile, many are evidently subpar at least temporarily when it comes to producing successful sperm (more than 10 percent of eggs do not hatch). Interestingly, the number of eggs failing to hatch is positively correlated with the number of females socially mated to a given male; a likely possibility is that with many females to fertilize, the sperm production of some males is unable to keep up with demand. As a result, females may "demand" to copulate with other males. (The problem of sperm depletion is not altogether unknown to human beings: Couples complaining of infertility are typically counseled to *reduce* their frequency of intercourse, since an overactive sex life can diminish a man's sperm count to the degree that it interferes with fertility.)

Strangely, female red-wings do not seem to prefer better-quality males; at least, they do not solicit EPCs from males that had the highest reproductive success the previous year. There is still much for us to learn about this business!

The red-winged blackbird population described above was studied in central Washington State. Although their situation does not seem unique, it may be misleading to generalize too readily to other populations, not to mention extrapolating to other species. Thus, although Washington State female red-wings initiate EPCs, in another closely monitored population, in New York State, female red-wings only accept or resist the advances of males. They do not initiate.

Although it seems likely that natural selection would provide males with the ability to make enough sperm, it is nonetheless possible that EPCs provide females with a kind of "sperm security," just in case their mate's supply is running low. Extra-pair copulations are frequent among house sparrows, and especially so among those that produce infertile eggs, suggesting that when females are mated to infertile males, they are more likely to seek EPCs. At this point, it isn't at all clear how a female can ascertain the sperm supply of her male partner.

It is possible that they can't, in which case it may pay them to mate with more than one male, just for insurance. Among Gunnison's prairie dogs, female do just that. It turns out that the rare Gunnison's female who copulates with only one male has a 92 percent probability of becoming pregnant and giving birth, while the probability is 100 percent for females that copulate with three or more different males. In addition, litter size is larger when the number of sexual partners is higher. Not all mammals show this pattern, however, including even some other ground squirrels, which are closely related to prairie dogs. In fact, even in another prairie dog species, the black-tailed prairie dog, there is no such tendency. In some mammals, there is actually a *reduction* in female reproductive success associated with mating with a second male, at least under laboratory conditions. But maybe this last finding isn't so contradictory after all: It makes sense that a female might fail to conceive in the presence of more than one male, whereas she might do just fine if she had the chance to consort with one male at a time, privately, and on her own terms.

Throughout most of the animal world, females have something that males want: their eggs. And nearly always, males are quite willing—even eager—to provide sperm. As a result, females are unlikely to be so desperate for sperm donors that they cannot exercise a degree of choice. This is in fact a useful way to look at the phenomenon of EPCs: They provide females with additional opportunities for choice, selecting a *genetic* partner independent—if need be—of their *social* partner.

Not surprisingly, females are less likely to face the question of "to fertilize or not to fertilize" than "to fertilize well, or not." According to legend, Cleopatra—not known for sexual abstention—was killed by an adder, a species of snake, which, it now appears, is no more monogamous than the famous queen herself. Thus, in a paper titled "Why Do Female Adders Copulate So Frequently?", a group of researchers reported that in at least one kind of adder, females that mate with several different males have fewer stillborn offspring than do their unlucky counterparts who are forced to mate with one or just a few partners. In such cases, the key contribution distinguishing one male from another seems to be whether his genes lead to healthy development after fertilization. There is a lot that can go wrong in the journey from fertilized egg to fully formed little snakelet. The more males a female adder mates with, the more likely she is to encounter a partner whose sperm will complement her own, leading to increased success in successfully negotiating the pitfalls of snake embryology.

This is as good a place as any to confront a possible misunderstanding, one that might otherwise bedevil many readers, here and elsewhere. Thus, some of you may balk at the notion of animals choosing their sexual partners with such exquisite care. You may feel tempted to throw this book across the room, exclaiming in frustration: "What are you talking about? How could female adders possibly know so much—indeed, anything at all!—about the precise pitfalls of their own embryology?" As a matter of fact, even well-trained biologists don't know a whole lot about snake embryology. The point is that living things have evolved the ability to engage in all sorts of activities without necessarily having any detailed understanding whatever about what they are doing or why. Flowers bloom in the spring without "knowing" that their seeds will be most successful as a result, because they will germinate in the summer. Animals—including human beings—engage in an extraordinary array of fancy biochemical, molecular, and electrical events (think of digestion, respiration, DNA transcription and translation, the immune response, indeed, even thinking itself) without consciously understanding either the general processes or the details.

So please don't get "hung up" on such details yourself, at least as they pertain to the ability of living things to behave adaptively, just as they grow and metabolize adaptively even if they have never read a textbook, attended a lecture, or conducted a single laboratory experiment!

We hope you can grant, then, that a female's motivation in obtaining one or more EPCs might well go beyond fertility insurance to include an unconscious, evolutionarily based search for "complementary" genes. But what, specifically, might such a female be seeking? Based on theory alone, it would seem that copulating with more than one male would not convey a benefit in terms of added genetic diversity, since part of the charm of gamete production is the generation of astronomical genetic diversity; the genetic variety per se within just one male's ejaculation is immense, offering nearly as much range as can be obtained by mating with several different males.

On the other hand, there are often substantial costs associated with inbreeding, the mating of close relatives. Physical deformities are common, for example, among the offspring of closely related European sand lizards. DNA fingerprinting has shown that when a female sand lizard mates with a closely related male, he is likely to sire a small proportion of the offspring produced; more distantly related ("outbred") males father a larger proportion of the young. By multiple mating, a questing female seems likely to increase the chance that she will find a partner who is genetically different from herself.

This is not just a reptile thing. In a bird species wonderfully called the splendid fairy-wren, a high frequency of EPCs is apparently due at least in part to the benefits of avoiding inbreeding. And mammals are not immune to such considerations. A group of genes known as the major histocompatability complex (MHC) serves as a key marker by which the immune system distinguishes "self" from "other." It also serves to indicate genetic closeness and is important in producing viable offspring. Among mice, offspring with incompatible MHC genes are spontaneously aborted. Interestingly, when female mice find themselves occupying the territories of males whose MHC genes are incompatible with their own, these females engage in EPCs with males from adjoining territories, whose MHC genes are a better match. (It appears that different MHC genes produce different odors, to which the females are sensitive.)

Female primates, for their part, often show a particular interest in mating with a male who is the new guy on the block. Paradoxically, although these new arrivals are nearly always low in social rank, they are often sexually appealing to females. For example, in one troop of Japanese macaques, a newly arrived male occupied the lowest social rank but mated with more different females than any other male. In one remarkable case, a female red howler monkey consistently turned down the local boys but was receptive any time she encountered a male from a neighboring troop.

The human equivalent—if any—isn't clear, but one hint may come from the interest and even fascination often generated by the "drifter," the mysterious newcomer. Even the cliché "You will meet a tall, dark stranger" might also capture some of the (literal) romance of novelty. On an evolutionary level, it is at least possible that this "strange-male preference" derives from inbreeding avoidance. In which case, it isn't so strange after all.

A key summary point is that in a wide range of species, females exercise direct choice as to their sexual partners, often choosing more than one. Although such behavior may prove risky for females if they are punished by their cuckolded male consort, when successful they could be repaid by bestowing the benefits of outbreeding upon their offspring.

Don't be too quick, however, to conclude that EPCs always pay for themselves via outbreeding. There are, it appears, different strokes for different species: Just as there is a downside to too much inbreeding, excessive outbreeding, too, can carry costs. In at least one case, EPCs seem, paradoxically, to be a mechanism for keeping genes *in* the family instead of introducing new ones. In one bird species, the pied flycatcher, breeding pairs that are genetically quite different are more likely to have extra-pair young in their nests than are those who are genetically more similar. In this species, therefore, it appears that females are prone to mitigate the effects of extreme outbreeding, seeking EPCs with males who are somewhat more genetically *similar* to themselves. The problem with excessive outbreeding is that it might break up locally adapted gene combinations, which simply means that by combining individuals who are too different, the resulting offspring might fall between two stools, landing on neither. Females may well choose as mates those likely to meet the Goldilocks criterion: a partner who is not too similar, and not too different, but Just Right. (It is also possible, incidentally, that when a male and female pied flycatcher are just "too different" genetically, they are somewhat behaviorally incompatible as well, which would lead directly to more EPCs.)

So far, in examining the female search for sexual variety, we have looked at simple genetic success versus failure: producing offspring or failing to do so. This barely scratches the surface when it comes to reasons why females may elect to mate with more than one male. There are many aspects to mating "well."

If females copulate with many different males, then in theory they can choose among the sperm of these various males and decide which one to favor with an egg or two, or, like an expert financial planner, they might even choose to diversify their genetic portfolio, allowing a preferred mix of different males to fertilize different numbers of their eggs—maybe even allotting certain eggs to certain sperm.

This may seem far-fetched, but it is not impossible. More plausible yet are a variety of EPC-related tactics by which females enhance the likelihood that their eggs will be combined with the best possible male genes. Imagine, for example, a female mated to a male via a long-term pair-bond. Imagine, further, that the male in question is something less than a sterling specimen: adequate, but nothing to write home about. Given the opportunity, in fact, the female would have chosen someone else. But since the species is generally "monogamous," she never had much opportunity to choose. After all, for every female there is, on average, one male, and vice versa. In a polygynous species, one male might be mated to a dozen or so females, in which case each of those 12 females might have been able to avail herself of the 1 in 12 males who is especially desirable, resulting in 12 happy females while also leaving on average 11 bachelor males. In such a case, their loss is the females' gain, since each harem-member has gotten an unusually high-quality mate—albeit the same one—who will presumably provide high-quality genes.

But in the case of our hypothetical monogamous situation, females have much less opportunity to choose a really classy male, since all the best ones have been taken, presumably by the best—most desirable—females. Furthermore, let's imagine that our monogamous female isn't such a prize-winner herself, so she wasn't exactly free to choose the male of her dreams. She had to "settle." Imagine, further, that her mate's sperm is good enough to fertilize all her eggs and to produce viable offspring. Still, it is one thing to be viable, another to be a raging success.

In order to reproduce at all, our female needed a social mate; otherwise, she wouldn't have a nest, for example, or a feeding territory, or the assistance of an adult male, necessary perhaps to defend her and her young or maybe to help provision the offspring. But recall that in obtaining these prerequisites—the material necessities of reproduction—she had to accept a male whose genetic traits are less than prepossessing. In evolutionary terms, it is not important that such a female may be "disappointed" or "dissatisfied" with her mate, except insofar as "disappointment" or "dissatisfaction" may be a human way of saying that she might be tempted to improve the genetic characteristics of her offspring by mating with one or more other males . . . likely one of those desirable hunks with whom she was unable to establish a pair-bond.

It is noteworthy that female animals only rarely have affairs with bachelors (who, after all, are likely to be rejects). Instead, they choose someone
else's mate, probably because he offers better genes, plus—as we shall see in
a bit—maybe other resources. Even harem-living females can sometimes
exercise some control over their genetic partners. In species ranging from
elephants to elephant seals, females seem to go out of their way to associate
with a dominant male. They also vocalize loudly when a subordinate male
attempts to mount them; this alerts other males to the copulatory attempt,
whereupon the most dominant male is likely to drive away the subordinate
interloper and mount the female himself. We cannot say whether the female
"knows what she is doing," but it seems clear that, as a result, the lady
elephant seal is more likely to be inseminated by a dominant male than by
a subordinate.

Rather than directly granting sexual favors to one male in preference to
others, female mate choice can thus be indirect. This might help explain
why, in many species, females are inclined to aggregate at a mating site, from
which a single dominant male can exclude other males. Or females can
advertise, for example, that they are sexually receptive, thus generating conditions that provoke males to compete among themselves. The conspicuous
estrous swellings of female primates may similarly have evolved in the serv-
ice of indirect mate choice, with the Technicolor posteriors and yummy
odors of estrous females inciting male–male competition, to the ultimate
genetic benefit of the females.

Female garter snakes engage in a kind of *coitus interruptus* that apparently enables them to control who fertilizes their precious eggs. In eight of
twelve observed copulations with unsuitable partners, females were seen to
rotate their bodies wildly, interrupting the mating and preventing the formation of a copulatory plug. Dominant male red jungle fowl roosters (the
wild forerunners of today's barnyard chickens) do not have to force females
to copulate with them; only subordinate roosters must stoop to such behavior. Interestingly, of ten such forced copulations that were observed in one
study, four were followed by vigorous feather shaking on the part of the
female, which resulted in sperm being ejected from the cloaca.

Here is an anecdotal account of something equally thought-provoking in
a common bird, the yellow warbler. It seems that on one occasion a particular male remained unmated throughout an entire breeding season, adopt-
ing instead the habit of forcing EPCs on "married" females. Once, after he
forced such a copulation, his victim immediately flew to her mate and successfully solicited a copulation with him! In this case, a parsimonious interpretation is that the victimized female yellow warbler "preferred" to have
her eggs fertilized by her social mate.

Even without pushy males, females are often confronted with another sexual difficulty: a sampling problem. After all, they generally encounter males sequentially; that is, one at a time, perhaps with a fairly long pause in between. With each male they meet, they must "decide" (either consciously or not) whether to mate or wait, all the while not "knowing" whether they will encounter any others. Moreover, our female must somehow evaluate the breeding quality of each male, presumably either comparing him with some internal standard or against her memory of other males already encountered. One option—although not the only one—is to mate with the first male to come along, then mate again only if any subsequent swain shows himself to be better than the previous one.

It may be costly or even impossible for females to sample many different males before settling on a mate. For example, a female pied flycatcher visits on average only 3.8 males before choosing one. When females compete vigorously among themselves for a limited number of especially desirable males, their options may be even more restricted. As a result, females can hardly be expected to make a very informed choice; or, at least, their "choice" of a mate may be largely a matter of settling for whatever they can get. However we look at it, depending on how many additional males they encounter after they are paired up, some females may be likely to discover a male who is more desirable than the one with whom they find themselves. Under these conditions, "till death do us part" does not make a whole lot of sense. More likely is a strategy of "having your cake and eating it, too."

Think of it as a kind of one-way ratchet, whereby females, after accepting an initial mating, will mate again, but only if, by doing so, they are "ratcheting up," improving the genetic situation of their offspring. The tactic would be to mate with a seemingly good male—one who meets the minimum criteria of being of the right species, the right sex, and basically adequate—then remain available to mate with a better one, if he shows up. In a species of salamander, the European smooth newts, females pick up the sperm packets of males with particularly large head crests. In one experiment, females were exposed to males varying in the size of their crests, separated by 20 days. The females mated the first time, then had the choice of remating a second time or continuing to lay eggs fertilized by the sperm of their first mate. In this situation, most females only remated if the male to whom they were exposed the second time had a larger crest than the one whose sperm they had initially accepted. (Crestfallen, in such cases, is a severe condition indeed.)

In at least one species of spider, females that have already mated are willing to remate if they encounter a male with body size and fighting ability superior to that of their previous mate. Not only that, but the offspring of such multiple-mating females have a higher growth rate than those produced by single-mating females.

For any of several reasons, females can find themselves paired with males who are not genetically the best: As already mentioned, if the species is socially monogamous, only one female gets the best male. Everyone else is "settling." In territorial species, a female generally chooses a male based on the quality of his territory, although she may also use other criteria: whether he provides good parental care, or is especially adept at foraging, or is skilled at defending their young from predators. Alternatively, a female may simply settle on familiar real estate, taking the male who is there. Biologists have tended to think that life is a package deal: By getting, say, a resource-rich male, a female also gets the best genes. But this need not always be true. If a male who is genetically subpar genetically ends up with a high-quality piece of real estate, he may also end up with a female who looks elsewhere when it comes to a sexual partner.

In nonterritorial species, where mates are chosen not for their resources but more often for their personal qualities, including their genetic attributes, EPCs may be less important. For example, among waterfowl such as ducks, pairing is based on individual traits, not on possession of real estate, and, significantly, female ducks are notable for how vigorously they resist attempted EPCs. It seems likely that in the duck world, most females are satisfied with the genetic makeup of their mates; hence, they are less inclined to copulate with anyone else.

In most vertebrates at least, females can control the timing of copulation, which in turn means that they can control fertilizations. The most common pattern in birds, for example, is for females to call the sexual shots: They initiate copulations and determine when to refrain. Surprisingly, perhaps, they stop copulating while the female is still fertile! At this time, they have obtained enough sperm to insure fertilization of their eggs but appear to be hedging their genetic bets, giving themselves the opportunity of "ratcheting up": If a better-quality male comes along, such a female has the option of copulating with him, too. Because of the "last male advantage" (last in, first out), this male is likely to fertilize more than his share. On the other hand, if no such desirable stud shows up, these females have lost nothing; their eggs will still be fertilized, this time by their social partner.

Another option—and probably a simpler one—is for the female to retain a memory of her last mate and choose a different one each time. This is the tactic followed by a strange little invertebrate known as a pseudoscorpion. In a well-designed research study, once-mated female pseudoscorpions were given the opportunity to mate with their earlier partner or a new male. After an interval of $1^1/2$ hours, females invariably preferred new males, rejecting their previous mates; after 48 hours, on the other hand, they were equally likely to mate with their old lovers or with new ones. (The males were equally willing—indeed, eager—regardless of the interval.) This seems to be a way for females to increase their chance of acquiring a diverse array of sperm, while also making sure that they get enough to fertilize their eggs. It had already been demonstrated that pseudoscorpion females—like female European adders—that mated with more than one male have more reproductive success than do females artificially restricted to mating with just one (that is, enforced monogamy). The key seems to be genetic incompatibility between some males and females rather than the intrinsic merit of a given male's sperm. It is less a matter of higher-versus lower-quality males than the "genetic fit" between any two would-be parents: The same male pseudoscorpion may have highly successful offspring with one female but many "stillborn" offspring with a different female.

But in these animals, males don't appear to differ outwardly, even though each is genetically distinct. So females apparently cannot determine whether a given male is the one for them. Neither are pseudoscorpion females blessed with an especially good memory. Under these conditions, it may be that the best strategy for lady pseudoscorpions is to cast their reproductive nets widely, mating with new males rather than old partners and thus making it likely that at least one of their sexual consorts will provide the right, matching sperm. (Pseudoscorpions are wonderful little creatures, by the way. They reside under the bark of decaying tropical trees, and in order to leave one such tree and make their way to another they must hitch a ride underneath the wing covers of another small invertebrate, the harlequin beetle. Male pseudoscorpions compete with each other to monopolize the limited travel space; no carry-on luggage permitted.)

In one type of beetle, the female remates more readily with a new male than with her previous partner, which suggests that she, too, is looking for genetic benefits associated, in all likelihood, with enhanced variety. In some insects, the female deposits an identifying odor on the male, allowing her to discriminate against him when it comes to remating!

Most commonly, however, it appears that females choose males with "good genes," as shown by the fact that males with larger sexual ornaments produce more viable offspring (e.g., peacocks). But how to reconcile this with multiple mating by females? If females are choosing the best male, why mate with more than one? Recall that a female may simply not be able to settle down and raise a family with the male of her dreams; he may already be taken. In this case, EPCs provide the opportunity for females to bond with one male—and indeed to rear children with him—but to copulate with another, and thereby obtain his genes.

Another possibility is to mate with your social partner, but do so again—and yet again, perhaps—with a more preferred male, if he shows up. Such "ratcheting" is not female promiscuity, by the way, because these females are not being sexually indiscriminate. Far from it. They are carefully evaluating the merits of their potential sexual partners, seeking to trade up when possible. Other cases, like that of the pseudoscorpions, where females cannot evaluate the quality of males and simply go for novelty, are closer to promiscuity. But even here, there is method to their sexual meandering. Virgin pseudoscorpions, for example, invariably pick up the first sperm packet deposited by a male; after this, 88 percent reject his *next* sperm offering. It isn't that such females have suddenly become uninterested in mating, however: Exposed to a new male shortly afterward, they again pick up his first sperm packet . . . and then refuse his subsequent ones. They're just uninterested in mating with the same male twice!

The most frequent pattern reveals not so much a preference for diversity as a predilection for quality. Male house sparrows, for example, possess dark throat badges, indicators of their macho qualities—and female house sparrows are more likely to have EPCs with males sporting large throat badges than with those whose throat badges are less impressive.

Most of the time, moreover, females choose their EPC partners from among the ranks of married males—who are likely to be of higher quality—rather than mating on the side with bachelors, who are generally bachelors for good reasons (that is, they were unsuccessful in obtaining a mate in the first place because of being lower quality).

Not only that, but females tend to select as EPC partners those males who are in some way superior to their current mates. Among zebra finches, females evaluate the sexual desirability of a male by the brightness of his bill color; they solicit EPCs only from males whose bills are not only bright, but brighter than those of their current mates.

Many female birds get to choose their mates in order as they—the females—arrive on the breeding territories; earlier-arriving females should therefore get the most desirable males, and later-arriving ones should be more likely to seek out EPCs. This appears to be the case.

If genetic diversity were the reason for EPCs, then females of species that produce only one egg per year should attempt EPCs only in some years, not each year. (One offspring with a given male cannot be more diverse than one offspring with another male!) On the other hand, if females are seeking genetic quality, they should be motivated to obtain such quality each year.

A study of razorbills, an ocean-going bird that produces only one egg per year, found that the females' inclination for EPCs remained consistent year to year. This supports the hypothesis that they are seeking quality rather than diversity.

The prevalence of EPCs is highly variable, from zero in some species to nearly 100 percent in others. Even within any one species, there are typically differences between male and female participation in EPCs, the general pattern being a higher proportion of female than of male involvement. It is not uncommon, for example, for about 15 percent of females, but only 7 percent of males, to engage in EPCs. This harkens back to the earlier generalization by sociobiologists that males are more variable in their reproductive success than are females, since a small proportion of males "enjoy" reproductive success that goes beyond their own pair-bond. Presumably, those that do so are especially desirable to the females in question. The chances are that any female willing to take the risks of an EPC would be able to obtain it, whereas males generally yearn to be called, but only few are chosen.

An important consideration—for both males and females—is the breeding system itself, for example, whether there are other "helpers" available to assist with child care. Fairy-wrens of Australia often breed cooperatively, with all males contributing to the feeding and defense of the young. DNA fingerprinting shows that more than 75 percent of the offspring are actually fathered by males outside the group, who provide no parental care, and that 95 percent of all broods contain young sired by extra-group fathers. This is the highest incidence of animal cuckoldry of which we are aware. Female fairy-wrens are very much in charge of whom they mate with: They successfully avoid all EPC attempts initiated by outside males, only mating when they initiate and solicit. And they are highly selective. Of 68 out-of-group potential fathers identified in one study, just 3 fathered nearly 50 percent of all the extra-pair offspring. As a result of such choice (and, in all likelihood, the reason for it as well), sons produced via EPCs are highly successful in obtaining EPCs.

Fairy-wrens do not always form large breeding groups. In some cases, they mate in pairs; in these situations, the paired male has a much higher chance of being the biological father than does the dominant male in a communally breeding group. Paired males also contribute much more to rearing their offspring, as predicted by the fact that they are likely to be the fathers. In communal breeding groups, females get less help from any given male, but, in return, they gain freedom of reproductive choice, via EPCs. When female fairy-wrens have helper males available—the group situation—they are liberated to choose good genes from outside the group. (Incidentally, these helpers are generally sons of the female; hence they gain an indirect genetic payoff from their mother's breeding success—no matter who the father is—while the adult female likely gains by being sexually liberated to choose the best possible EPC partner.).

Not surprisingly, females of any species that are sitting pretty, mated to an especially well-endowed male, are less inclined to stray sexually, whereas those whose mates are less desirable are more likely to try one or more EPCs . . . and when they do so, to insist on "moving up," reserving their extracurricular mating for those males who offer a better genetic package than their current mate.

L et's assume that, at least in some cases, females are in fact choosing good genes. What, precisely, does this mean? How can one set of genes be better than another? In lots of ways.

For one, good genes could simply be those that lead to healthier offspring. Probably the best example of this comes from research conducted by Allison Welch, then a graduate student at the University of Missouri, Columbia, and her colleagues. They studied gray tree frogs, a species in which females prefer to mate with males whose songs are comparatively lengthy (about two seconds long) rather than very brief (about one second long). By acting on this preference, females get good genes, leading to offspring who are more fit. Welch and associates fertilized female eggs with sperm from long- and short-calling frogs, then compared the resulting offspring both as tadpoles and after they had metamorphosed into frogs. The key result: The offspring of long-callers fared better. The key interpretation: Long-calling male gray tree frogs produce offspring that are more likely to survive and, eventually, reproduce. Accordingly, females are well advised to mate with long-callers rather than short callers . . . which they do.

Many animals are brightly colored, and often the males are especially adorned. It has been suggested that bright coloration has evolved among males as a cue allowing females to choose as extra-pair partners those males that are especially healthy—parasite-free or parasite-resistant. The saga continues: Among yellowhammers—a species of European finch in which males are bright yellow and females are comparatively drab, and in which old, colorful males are especially successful in obtaining EPCs—the more brightly colored the male, the less likely he is to be infected by parasites. At the same time, the dullest, least-yellow males are most likely to be cuckolded. Among yellowhammers, the brightest, yellowest males are also the oldest, so degree of yellowness and brightness is a reliable indicator that one is carrying genes conducive to longevity.

Adding to the emerging picture of females engaging in EPCs so as to accrue "good genes" is the fact that species with a high frequency of EPCs are especially likely to be sexually dichromatic (substantial male–female differences in coloration) and to have large spleens for their body size: This suggests a proportionately more well-developed immune system. Such a correlation does not strictly prove anything, but it is consistent with the notion that females in such species seek EPCs in order to obtain greater disease and parasite resistance for their offspring.

An underlying assumption connecting these examples is that males that are especially healthy, and whose health is to some extent heritable, will be preferentially chosen by females as EPC partners. We might define an "attractive" male as one who is visited by many neighboring females, in search of EPCs, and an "unattractive" male as one who is not comparably visited and, thus, who is not in comparable demand. It is then also revealing that females associated with attractive males do not leave their male prospecting for EPCs, whereas females associated with "unattractive" males (that is, those who are not visited by other females) frequently visit their male neighbors.

How, one might ask, do females judge the quality of their mate? And, similarly, how do they judge the quality of their neighbors and possible EPC partners in comparison? A group of Dutch researchers recount one suggestive event: "A male [blue tit] injured one wing just before egg-laying. His female visited both neighbors and both shared paternity with the territorial male. After the young hatched, the territorial male died." All told, of the five males that died between 1 and 3 weeks after their female started incubation, four had a high proportion of extra-pair young in their nest ($^5/_6$, $^2/_{10}$, $^3/_3$, $^4/_7$). It is entirely possible that females use the condition of a male during the breeding season as a clue to indicate his quality and, in turn, use their male's quality as a key determiner of whether or not to seek EPCs.

In at least two other bird species—tree swallows and blue tits—females mated to males that are of comparatively poor quality are more actively involved in seeking EPCs than are females mated to good-quality males. What constitutes "good quality" in such cases is somewhat problematic. But we are beginning to get some hints. Among blue tits, for example, males differ in their probability of surviving the winter, and males that are less likely to make it during the coming winter are more likely to be cuckolded; similarly, males that do the cuckolding are more likely to overwinter successfully. How do female blue tits distinguish the winter-hardy from the winter-wimps? No one knows. (Yet.)

Even when females do not bestow EPCs on physically distinguished males, they may nonetheless be showing a preference for those likely to be of higher quality, perhaps because they are behaviorally distinguished . . . and, more often than not, genetically distinguished as well. Nothing succeeds, we are told, like success. And indeed, social success—measured by one's position in a dominance hierarchy—succeeds mightily when it comes to securing extra-pair copulations. (Maybe this is what Henry Kissinger meant when he noted that "power is the best aphrodisiac.")

There is a widespread tendency for females to prefer dominant males when it comes to bestowing their EPC favors. Dominant male cattle egrets and white ibis—who are successful in male–male fights—are particularly likely to obtain EPCs. Similarly, among black-capped chickadees (a close North American relative of the European blue tit and a common winter participant at bird-feeders), females reserve their EPCs for males whose dominance status is higher than that of their own mates. Over the course of a 14-year study of black-capped chickadees, ornithologist Susan M. Smith observed 13 apparently successful EPCs done by individuals who were color-banded and, thus, whose identities and social ranks were known. In all 13 cases, the female's EPC partner was higher-ranking than her social mate. No females mated to alpha males ever engaged in an EPC. Not only that, but EPCs are mostly solicited by females—individuals who presumably had to settle for less dominant mates and were trying to make up for this deficit. (There's that ratchet again.)

Chickadees, it should be noted, generally mate for life. Smith observed seven cases of divorce; in five of these, a lower-ranked female deserted her mate and established a new social and sexual alliance with a recently widowed alpha male. Social dominance often increases with age; in addition, older males—if only because they have survived so long—are obviously capable of longevity and may well carry genes that promote longer life. Accordingly, older male red-winged blackbirds are more successful in obtaining EPCs. A similar age-related pattern occurs in the European rook (a relative of the North American crow): Older paired males engage in EPCs with younger females. Could this be because younger females were likely to be paired with younger males . . . while preferring older ones?

Let us grant that in many different species, females often seek EPCs with males that are especially attractive and dominant (as well as suitably mature). Although this is probably due to a preference on the part of females for males with good genes, the natural world is tricky. Thus, it is possible that dominant and older males are simply more likely to be available for EPCs because they have more sperm to spare or because, as a result of their dominance, they are less likely to be excluded from the territories of other monogamous males. Although females may well seek—and get—good genes from such liaisons, this is not the same as guaranteeing that they engage in EPCs with high-quality males in order to obtain such genes.

Nonetheless, the evidence is accumulating and is increasingly persuasive.

In songbirds, a male's quality may itself be reflected in his singing. In the previous chapter, we encountered the European great reed warbler, among which females choose EPCs with males who have large song repertoires: As it happens, the survival of young reed warblers is positively correlated with the size of the genetic father's song repertoire. So there is a practical, immediate significance to more songs: better genes. And as one might expect, females of several species give especially intense copulation displays in response to hearing an elaborate song repertoire. (So perhaps there is something to the old tradition of serenading one's lady love.)

If females engage in EPCs with males who offer especially good genes, then an interesting—and controversial—possibility arises, suggested by the observation that in some species females resist EPCs, sometimes quite vigorously. The possibility is this: Females could gain an advantage for their offspring (good genes) if they make sure that their EPC partner really is of high quality by resisting males' EPC attempts, only submitting to one who shows himself to be unusually determined, competent, and—almost literally—irresistible. As a result, her male offspring might also likely be determined, competent, and comparably irresistible when it comes to obtaining EPCs themselves. A chip off the old block.

On the other hand, female resistance to EPCs, when it occurs, may be genuine: Sometimes no really does mean NO! On balance, in fact, female EPC resistance is probably more frequent than acquiescence or solicitation. (Given the obvious payoff to them, it is not surprising that males seek EPCs and that they typically do so more actively than females. The reason for examining female solicitation of and acquiescence in EPCs is that the phenomenon is so counterintuitive—and yet so frequent.)

Is there a simple, one-size-fits-all, easily discernible characteristic that might provide convenient information about whether an individual is carrying "good genes"? Maybe there is. The characteristic is symmetry, specifically the degree of left–right correspondence between the two sides of an individual's body, whether arms, legs, eyes, ears, wings, flippers, and so forth. All vertebrates are bilaterally symmetrical (jellyfish, sea urchins, and starfish, by contrast, are radially symmetrical). In the case of bilaterally symmetrical creatures, left and right are not controlled by different genes, so asymmetry—difference between left and right—is widely assumed to reflect some sort of developmental perturbations, whether caused by poor nutritional status, toxins, mutations, or pathogens.

It turns out that males with low asymmetry have high mating success and vice versa; males who are symmetrical are widely seen as attractive, and vice versa for those who are lopsided. This has been found for a variety of animals, from insects to primates. Two research papers dealing with barn swallows and published in the same year by noted Danish researcher Anders Møller tell an impressively logical tale. One was titled "Female Swallow Preference for Symmetrical Male Sexual Ornaments." In plain English: Female swallows prefer males whose forked tails are equal in length. The second study was titled "Parasites Differentially Increase the Degree of Fluctuating Asymmetry in Secondary Sexual Characters." In plain English: Males infested with parasites tend to be lopsided rather than symmetrical. Put the two together: Female swallows prefer males who are symmetrical, in all likelihood because such males are not parasite-laden.

As it happens, some of the most persuasive evidence for the role of asymmetry comes from studies of human beings. The procedure is surprisingly simple: Measure a number of body parts that are bilateral (such as feet, hands, ankles, wrists, elbows, ear length, and ear width), obtain a composite index of degree of symmetry (or asymmetry), and see if the resulting measure correlates with perceptions of physical attractiveness. It does: More symmetry equals better looking. Not only that, but symmetrical men generally have a relatively high number of *sexual* partners, so the judgment by women isn't merely theoretical! Women even report more orgasms when having sex with symmetrical men.

This leads to the prediction—especially relevant for our purposes—that symmetrical men will have a comparatively large number of EPC partners and vice versa. The prediction holds. A study of more than 200 college students asked them a number of questions, guaranteeing—for obvious reasons—to keep their responses anonymous. Specifically, they were asked about any sexual liaisons they had had (1) with someone who was already involved in a romantic relationship with someone else and (2) while they were themselves involved in a romantic relationship with someone else. In addition to being measured for physical symmetry, respondents were queried as to their age, socioeconomic status, likely future salary, and emotional attachment style. They were also photographed for independent assessment of their physical attractiveness.

Among the interesting findings: Symmetrical men reported more EPC partners—both when they were paired with someone else and when they were the "third person"—than did asymmetrical men, a result that persisted when any effects of social status, likely salary, age, and even physical attractiveness were eliminated. So, when it comes to already-paired females engaging in some sex on the side, barn swallows are not alone: Women, too, prefer to dally with members of the opposite sex who are symmetrical. Such a preference may well work the other way, too, although to date it has received less research attention: It is a good bet that men also prefer symmetrical women. The suggestion has even been made that part of the widespread male fascination with female breasts is that such protuberant, bilateral organs provide a good opportunity to assess symmetry! (By contrast, a penis would seem to offer much less opportunity for a connoisseur's assessment, since—whether dangling limply or standing proudly erect—it is nonetheless a lowly singleton, a mere midline member. Tough luck. But whose?)

You might have noticed at least two logical problems with all this. First, the information on numbers of EPCs was obtained by so-called self-reports, that is, what people say they did, as distinct from what they actually have done. This may be a serious, if unavoidable, problem. On the other hand, difficulty arises only if symmetrical (or asymmetrical) people are consistently prone to exaggerate (or, alternatively, understate) their frequency of sexual dalliances . . . situations that seem unlikely. A second potential problem is one of interpretation: Even if his physical symmetry genuinely correlates with a man's extra-pair copulations, it isn't clear, for example, whether women are attuned to the actual physical symmetry of potential sexual partners or whether symmetry correlates with something else (self-confidence, unknown pheromones, cosmic emanations, whatever).

In any event, it is also interesting to note that in the study just described, the number of a woman's out-of-pair partners correlated with her "emotional attachment style." Each subject (male and female) was given an "attachment index," based on two different styles: "avoidant attachment" or "anxious attachment." Avoidant attachment included agreement or disagreement with such statements as "I am nervous whenever anyone gets too close to me," while a typical sample item for the anxious-attachment scale would be "I often worry that my partner doesn't really love me." The results? Women with a higher level of anxious attachment had more out-of-pair lovers, whereas those with a higher level of avoidant attachment had fewer. A woman's degree of physical symmetry did not predict her number of out-of-pair partners.

These combined findings are consistent with the basic biology of male–female differences: Men's out-of-pair sex correlated with a physical trait that presumably says something about their desirability, whereas women's out-of-pair sex correlated with a mental trait that presumably says something about their willingness to have such a relationship. The implication is that men are generally willing, that women are generally able, and that women are most sexual with men who are symmetrical.

This much is clear: Females are inclined to have EPCs with males who have good genes. And as we have seen, "good genes" can include many things: being sufficiently different from the female in question (but not too different), being genetically complementary in other ways, or carrying health-related genes. But this isn't all. If certain characteristics (symmetry, bright plumage) indicate good genes and if, as a result, females are at an evolutionary advantage if they prefer these characteristics, then the stage is set for yet another wrinkle in the EPC saga: Females can benefit by preferring those males whose only virtue is that they are preferred by other females! Such a preference might well begin with traits that are "genuine," such as symmetry or bright plumage, but as the pioneering evolutionary geneticist R. A. Fisher pointed out decades ago, it could quickly develop a life of its own.

Indeed, there have been studies showing that in some cases female choice is driven by nothing other than female choice itself; that is, females sometimes choose mates not because they produce healthier or longer-lived offspring, but simply because those offspring—especially the sons—are themselves likely to be chosen by the next generation of females. This idea, now known as the "sexy son hypothesis," suggests that females may choose males simply because other females (a generation later) are likely to have the same preference. As a result, a female is well advised to be seduced by sexy males, even if these males are not exceptionally healthy or even likely to produce exceptionally healthy offspring, so long as the female's sons will be "sexy" . . . that is, attractive to the next generation of females. A kind of bandwagon effect.

For example, in a species of sandfly, females evince clear preference as to mates. In one experiment, females were denied the opportunity to exercise choice and were forced to mate with either preferred males or males who would otherwise be shunned. There was no impact of paternity on the overall health or viability of their offspring. But the offspring of preferred males were themselves preferred, just as shunned males produced sons who were shunned in turn.

When it comes to EPCs, females of many species are especially likely to mate with males who are more attractive than their partner. You can almost hear the females—whether already mated or not—spotting the animal equivalent of a movie star and sighing to themselves: "I want to have *his* kids." If so, the reason appears to be that, at an unconscious level, they can hear the echoes of other females saying the same thing about *their* future off-spring, thereby promising a larger number of grandchildren for the besot-ted, starstruck, would-be mother . . . who is now a candidate for one or more extra-pair copulations with the lucky hunk.

The converse also holds: Make a male less attractive, and his mate is more likely to look elsewhere for male genes. There is, for example, a small, strikingly colored socially monogamous Euroasian bird known as a bluethroat. Males have—not surprisingly—bright-blue throats; female throats are white. When researchers from the University of Oslo, in Norway, used dye to diminish the blueness of their mates' throats, female bluethroats were more likely to engage in EPCs. (It is also interesting that the de-blued males apparently perceived somehow that they were less attractive than before, perhaps because of changes in their mates' behavior, since they in-creased their mate-guarding activities, although to no avail.)

Female choice can also be influenced by what is popular or stylish at the moment. The phenomenon has been called "mate copying." Here's how it works: A female guppy is given a choice between two different males. This female then observes the male she had rejected being chosen by another female (actually, an artificial model of a female, manipulated by the experi-menters). Then the choice test is repeated, whereupon the female is likely to change her mind and prefer the male she had initially rejected but whom she had subsequently observed to be "popular." Not only that, but younger females are likely to copy the preferences shown by older females.

There is abundant evidence that sexy males get more EPCs, independent, perhaps, of whether they are really healthier or carrying genes that are "bet-ter" in any other sense. Male swallows whose forked tails are artificially lengthened obtain a mate 10 days earlier than normal males; they are eight times more likely to mate again and produce a second brood; and they are twice as likely to have one or more EPCs with an already-mated female. In one especially impressive study, three different kinds of male barn swallows were created: those whose tail forks were shortened, those whose tail forks were lengthened, and a control group whose tails were cut but then glued back together with no change. The results: Extra-pair offspring made up about 60 percent of the nestlings associated with tail-shortened males, as compared to 40 percent of controls and about 12 percent of males whose tails had been artificially elongated. At the same time, the number of bio-logical young reared in their nests increased directly with tail length.

Among house sparrows, long forked tails don't make a male sexy; large black throat patches do. Males engaging in EPCs are particularly likely to have impressive throat patches, and females are more likely to be involved in an EPC with a male whose black throat patch is larger than that of their "husband." A comparable finding applies to another type of bird, the zebra finch, among whom male attractiveness depends on the color of the beak (in the world of zebra finches, a red beak is "hot"). Researchers have even found that, in the case of zebra finches, the color of leg bands, installed by the experimenters, influences male attractiveness and, thus, females' penchant for EPCs. Again, red is desirable: Females mated to males sporting red bands are unlikely to mate outside the pair, while those mated to green-banded males are more likely to have "affairs," with the resulting young fathered by more appealing males.

In the preceding examples, if the females' goal was to increase the genetic diversity of their offspring, then all females should be equally inclined to EPCs. On the other hand, if increasing genetic *quality* or attractiveness is the goal, then females mated with particularly low-quality males should be especially EPC-prone (which they are). If there are only a few good males—and especially if these are very good—then EPCs should be particularly frequent. By contrast, if all males are pretty much so-so, EPCs would offer less genetic payoff. We can also predict that EPCs should be less frequent on islands—where genetic diversity is low—than on mainlands, where it is high. Similarly for populations that have recently gone through a genetic "bottleneck," resulting in less genetic diversity: Free-living cheetahs, for example, are notoriously inbred and lacking in genetic diversity; it seems likely that female cheetahs are not especially inclined to multiple mating or susceptible to EPCs.

E volution often works in strange and unexpected ways. This is certainly true when it comes to female preference for particular male traits. Thus, in addition to possibly preferring males who make enough sperm and who carry genes that are suitably varied, complementary, health-related, and sexy, there is good reason to think that females may even favor certain males based on characteristics of their sperm alone.

After all, not all sperm are created equal. Those from different males differ in the likelihood that they will succeed in fertilizing an egg. Studies have been conducted in which a wide range of animals—insects, chickens, mice, rabbits, pigs, cattle—were inseminated with similar amounts of sperm from two or three males. The findings are that, in nearly every case, sperm from a particular male are far more effective than those of other males. There are many reasons for this: Sperm come in different shapes and sizes and with different metabolisms, swimming abilities, and chemical characteristics—all of which can influence sperm motility, longevity, ability to penetrate the egg, ability to survive within the female genital tract, and so forth.

Since males differ in their capacity to fertilize eggs, it is quite reasonable that females would set up situations that exaggerate these differences. Why? Because it would benefit females to be fertilized by high-fertilizing males, since their own male offspring would then be likely to be high-fertilizers themselves. (We have already encountered the "sexy son hypothesis." Now, meet its close relative, the "sexy sperm hypothesis.")

The simplest way for females to ensure that they are fertilized by sperm that are sexy—or, at least, successful in sperm–sperm competition—is for the females to set up competitions; that is, to mate with more than one male. And this is precisely what many females do. Columbian ground squirrel females, for example, are in estrus for only about four hours per year, yet during this brief window of sexual opportunity, female ground squirrels are very busy, copulating with an average of 4.4 males. In another species of ground squirrel, females are in estrus for less than seven hours per year, and yet during this period, they copulate with an average of 6 to 7 males. And there is no reason to think that ground squirrel females are especially sexually profligate: Even when monogamy is not at issue, multiple mating by females is widespread, and sperm competition may well be the major reason.

But the story is far from over. If females profit by staging sperm competition, it stands to reason that they would profit even more if they could make the competition especially intense by creating a genital environment that is challenging and difficult, not only consisting of many different participants but also outfitted with obstacles both mechanical and physiological. For example, by extending the length of one's genital tract, the swimming capacity of sperm is highlighted. Insects and spiders are especially notorious for having tortuous reproductive systems, a series of lengthy gauntlets through which sperm must swim. May the best man— rather, the best sperm—win. (The resulting pattern has been called "cryptic female choice," something likely to be especially important in cases when females have relatively little opportunity for choosing a mating partner more overtly.)

Incidentally, there is nothing unique about females setting high standards for would-be suitors: defend a territory, engage in suitable (and often difficult) courtship rituals, provide lots of food during and after courtship, win a jousting tournament, slay a dragon, swim the Hellespont, and so forth. Consider, for example, that chimpanzee females commonly mate with many different males. It is probably no coincidence that male chimpanzees, in addition to sporting exceptionally large testicles (enabling them to produce large quantities of sperm), also have long penises. A longer penis almost certainly gives its possessor's sperm an advantage, since they could be deposited closer to the cervix. This, in turn, would be especially significant if sperm from more than one male are competing within a female, which of course is precisely what happens if the female mates with multiple males while she is ovulating. (And this is exactly what female chimps do.)

In one study, detailed measurements of 11 males and 19 females showed that in 10 of the 11 males, penis length actually exceeded the vaginal depth of 14 of the 19 females! This is not as strange as it might appear, since the length of a chimpanzee's vagina varies with her menstrual cycle, becoming maximum when the pink sexual skin is maximally swollen, which is also at the time of ovulation. Among some females, vaginal length increases during ovulation by as much as 50 percent. As a result, 7 of the 11 males described above would have been unable to reach the cervix of *any* of the 19 females, while the remaining 4 would only be capable of reaching some of them. By increasing the length of their vaginas when they ovulate, female chimpanzees make it more difficult for males to fertilize them, giving an advantage to sexual partners who produce sperm that are especially abundant, mobile, capable of withstanding the rigors of the vaginal environment, and delivered by a long penis. (Is this *why* the vaginas of ovulating chimps grow longer? Good question.)

Strangely enough, among all species practicing internal fertilization, it is virtually unknown for the male to introduce his sperm directly upon the female's eggs, which would result in immediate fertilization. Always the eggs are kept back, deeper inside, while the sperm are deposited in some sort of antechamber, from which they must proceed along various twisty canals, sluiceways, and storage containers, often lined with cilia (which not uncommonly beat in the opposite direction), traversing what—for a tiny sperm— must seem like hundreds of miles, generally through rather inhospitable terrain, chemically debilitating if not outright lethal, and, as if that were not enough, frequently patrolled by comparatively huge and aggressive sperm-eating phagocytes.

In an article titled "Why Do Females Make It So Difficult for Males to Fertilize Their Eggs?", biologists Tim Birkhead, Anders Møller, and W. J. Sutherland emphasized that "in terms of its structure, chemical composition and immune response, the female reproductive tract of mammals and birds is particularly hostile to sperm" and that, as a result, females are likely to be fertilized by the fittest sperm or, at least, to minimize the risk that they will be fertilized by the "worst." Imagine that males differ in their ability to overcome such female hostility. A female who indiscriminately allows any male to fertilize her eggs is likely to produce sons who themselves will only be able to fertilize a small subset of the females in their generation. By contrast, by setting up a difficult and demanding obstacle course, a female makes it likely that her sons will be able to fertilize most females.

If it were in the interests of females to make fertilization easy, it would not require much of a stretch in evolutionary potential to imagine a very different situation, in which sperm were ejaculated onto—or, at worst, very near—the eggs they were "intended" to fertilize. But although it is assuredly in the interests of females for fertilization to occur (just as it is in the interest of males), there is very little reason for females to make it easy. Indeed, there are good reasons to make it difficult. Or rather, one good reason: sperm competition.

"The phrase sperm competition," according to entomologist and evolutionary theorist William Eberhard, "evokes the image of armies of tiny, one-tailed soldiers racing up female ducts, or struggling in hand-to-hand combat to gain access to large, passive treasures. It emphasizes, as has been traditional in biology, the active male role in male–female interactions."

But in fact, as Eberhard points out, females may well hold the key to victory in sperm competition:

> Simple, relatively insignificant movements of the female, such as a few "downstream" peristaltic twitches of her reproductive tract, or beating in the wrong direction of the cilia lining the reproductive tract, can result in the army of little "warriors" being unceremoniously dumped from her body, or being diverted to other internal sites where they will be digested.

Responding, in all likelihood, to female maneuvering, male insects, for example, engage in a wide array of peculiar antics during copulation. William Eberhard, one last time: "Males of different insect species lick, tap, rub, push, kick, stroke, shake, squeeze, feed, sing to, and vibrate the female during copulation." All this, it seems, in an effort to overcome the female's resistance.

Females set up numerous barriers to fertilization; for example, the vaginas of most mammals have a very low pH, which is detrimental to the survival of sperm. The traditional explanation for an acidic vagina is that it reduces the danger of microbial infection. Supposedly, therefore, females do what they have to do—that is, acidify their reproductive tracts—to protect themselves, leaving sperm to cope however they can. Once females have opted to lower their vaginal pH, males must attempt to deliver sperm that will make the best of their difficult situation. With lots of sperm being produced, at least some have a good chance of making it. (And after all, any males that elected to opt out of the competition would be truly left out, supplanted by those that took the plunge, however acidic.)

At the same time—and here is the new twist—females are free to lower their pH even more, giving rise to additional male–male competition, resulting perhaps in acid-resistant, armored gladiatorial sperm whose voyaging capacity might rival that of Marco Polo, seminal secretions that temporarily increase pH, and so forth. Either way, females could in this manner end up using their low vaginal pH as a kind of sperm-screening device.

The female reproductive tract is notoriously unfriendly to sperm in other respects. Not uncommonly, the cervix is home to millions of leukocytes (white blood cells) that in other contexts devour bacterial invaders but that have quite an appetite for invading sperm as well; after all, sperm, too, are foreign to the female's body. Among human beings in particular, there are also high concentrations of antisperm antibodies in the cervical mucus. And to make things harder yet, female reproductive anatomy is typically elaborate, with eggs reachable only after sperm have completed a lengthy, twisty, and difficult upstream migration. Even then, sperm do not typically receive a warm hero's welcome: Eggs themselves are often stubbornly difficult to penetrate.

The effectiveness of these various antisperm defenses is attested by the fact that only a minute proportion of the vast numbers of sperm introduced into a female even get near her eggs. Until recently, this has widely been seen as due to the vagaries of chance, but the point is that it may not be a matter of chance at all; rather, antifertilization barriers may be specifically erected by females. It would be especially interesting to see if species characterized by a high level of EPCs turn out to be those in which female reproductive tracts are especially tortuous or possess a high level of antisperm antibodies, a particularly low vaginal pH, eggs that are unusually difficult to penetrate, and so forth.

At the same time, researchers must beware a tendency to overinterpret the mating advantages enjoyed by certain males over others, seeing any such imbalance as evidence that females are necessarily promoting sperm competition when they might not be doing anything of the sort. For example, it was found that when female dungflies copulate with large and small males, the former fertilize the lion's share of the eggs. Tempting as it is to attribute this to active selection by females of the sperm of large males, it turns out that this large-male advantage is simply due to the fact that large males transfer sperm at a higher rate. Nonetheless, it seems increasingly clear that in many cases—perhaps most—females are active participants in their own fertilization, not just in the choice of whom they mate with but also regarding what happens afterward.

If a female has only one sexual partner, there is no payoff for her in making things difficult for him. But if she mates with several males, it would probably be in a female's interest for her eggs to be "hard to get"— guarded, like Sleeping Beauty, by dragons, impenetrable thornbushes, and other daunting barriers. The successful Prince Charming should be not only charming, but also persevering and capable of producing sperm that are equally so.

But at the same time, females would be ill advised to carry this too far. Fussy is fine, but those who set the bar so high that no male could succeed in fertilizing their eggs would be strongly selected against, outcompeted by those whose expectations are more modest and realistic. Therefore, even the most well-defended egg-bearers are likely to be penetrated eventually—and when they are, it is likely to be by the most well-endowed males, possibly with a little help from the females themselves.

We might call it the Atalanta solution, after the Greek story of Atalanta and Hippomenes. According to master myth-recounter Thomas Bulfinch, Atalanta was as fleet-footed as she was lovely, and quite a catch in more ways than one:

> To all suitors (for she had many) she imposed a condition which was generally effectual in relieving her of their persecutions—"I will be the prize of him who shall conquer me in a race; but death must be the penalty of all who try and fail." In spite of this hard condition some would try. Hippomenes was to be judge of the race. "Can it be possible that any will be so rash as to risk so much for a wife?" said he. But when he saw her lay aside her robe for the race, he changed his mind, and said, "Pardon me, youths, I knew not the prize you were competing for." As he surveyed them he wished them all to be beaten and swelled with envy of any one that seemed at all likely to win. While such were his thoughts, the virgin darted forward. As she ran she looked more beautiful than ever. The breezes seemed to give wings to her feet; her hair flew over her shoulders. . . . All her competitors were distanced, and were put to death without mercy. Hippomenes, not daunted by this result, fixing his eyes on the virgin, said, "Why boast of beating those laggards? I offer myself for the contest."
>
> Atalanta looked at him with a pitying countenance, and hardly knew whether she would rather conquer him or not. "What god can tempt one so young and handsome to throw himself away? I pity him, not for his beauty (yet he is beautiful), but for his youth. I wish he would give up the race, or if he will be so mad, I hope he may outrun me."

Hippomenes could not outrun Atalanta, but he got some help from Venus. The goddess of love obligingly provided Hippomenes with three golden apples, which, at strategic points during the race, he threw down. And Atalanta, equally obliging, stopped to pick them up. As a result, Hippomenes won the race—and Atalanta.


In this chapter, we have sampled part of the female perspective on extra-pair copulations, concentrating on the EPC partner's genes. Next, we continue our female focus, turning to some other considerations.

# CHAPTER FOUR

# Undermining the Myth: Females (Other Considerations)

"Seek simplicity," advised Alfred North Whitehead, "and then distrust it." In the previous chapter, we sought to understand why females engage in EPCs by looking simply at female choice of male genes. (In the process, we found that even this "one" explanation is a far cry from simplicity, connected as it is to matters of fertilization, health, diversity, dominance, desirability, "sexy sons," and sperm competition—the latter perhaps engineered by females.) Now we turn to other factors, each of which appears to offer the seductive Siren call of simplicity, although in reality each is complex and subtly entwined with many others. Such understanding is a consummation devoutly to be sought . . . and then, perhaps, distrusted.

When done by males, EPCs are indeed interesting but, in a sense, unremarkable, since they are a logical, "simple" extension of the expected male sexual strategy: Given what it means to be male (i.e., a producer of comparatively cheap, easily replaced sperm), we can expect males, even if already mated, to be "fast and loose," and undeterred even if their extrapair partner is also already mated. If males are successful in garnering additional copulations, and if these copulations are with fertile females, they generate their own evolutionary reward, because any gene-influenced tendency to engage in EPCs will be promoted into the next generation. In short, an EPC is likely to be a male success or, more precisely, a success for the EPC-inclined genes residing within such males.

At the same time, from the perspective of the female involved—and assuming said female gains in the process—a departure from monogamy is just as validly a female strategy. Females, no less than males, have much to gain from EPCs. Biologists, too, have much to gain by acknowledging this, and recently they have begun to do so. There has been, in fact, a substantial change in focus on the part of many biologists, and the result has been a more nuanced and accurate sense of how living things go about their lives. Instead of limiting themselves to the male's perspective and identifying animal social systems as polygyny, monogamy, or monogamy plus adultery ("mixed male strategies"), researchers are talking more these days about polyandry: one female mating with many males, regardless of the apparent social system.

In the past, the word *polyandry* was restricted to those exceedingly rare cases in which a single female is simultaneously bonded to more than one male. But, increasingly, we see that a domestic system that is ostensibly monogamous (one male/one female) or even polygynous (one male/many females) may be genetically polyandrous, with the female mating with more than one male, regardless of what the social system appears to be.

At the risk of overdosing on terminology, it is time to employ a new term, increasingly used by biologists: *monandrous* (literally "one male"), to be distinguished from *polyandrous* ("many males"). It has the virtue of introducing—finally—a female-centered perspective into the study of mating systems. Using the word *polygynous*, for example, to describe the sexual life of a harem-forming species may be useful in presenting the perspective of the male: *Polygynous* means many females, and a successful polygynous male gets to mate with many females. But what about the females' viewpoint? Calling them *polygynous* implies that they are limited to mating with one male, the harem-keeper. In fact, we used to think that polygyny *meant* multiple mating for the male, single mating for the females. But not any longer. Polygynously mated females can be faithful to the one harem-keeper, in which case they are also *monandrous* ("one male"). But now we know that even in harem-dwelling species, females sometimes have additional EPCs with outside males, in which case they are polyandrous as well as polygynously mated.

So, it is useful to employ some new words, reflecting our new understanding. *Monandrous* ("one male") means that a female mates with just one male. Thus, a socially monogamous female may also be monandrous, in which case her sexual life matches her social life—she mates only with one male. Or she might combine social monogamy with genetic polyandry, having EPCs with more than one male. Similarly, a polygynous (harem-dwelling) female may be monandrous—mating only with one male, presumably the harem-keeper—or she might be both polygynous and polyandrous—socially part of one male's harem but likely to be involved sexually with one or more additional males, via EPCs.

Note to Lord Whitehead: It is one thing to seek simplicity, quite another to find it. (Distrusting the results is yet another matter!)

Why do females copulate so frequently? After all, it seems clear that females don't have to be inseminated very often just to get their eggs fertilized. Since eggs are much larger than sperm, they are necessarily produced in much smaller numbers. Just a handful of matings should therefore suffice. In a now-classic and much-cited study published in 1948, geneticist A. J. Bateman placed a limited number of male and female fruit flies in small containers and then observed their sexual behavior as well as the reproductive outcome. Obviously, zero copulations resulted in zero reproduction, for either sex. And the breeding payoff of a single copulation was pretty much the same for males and females (although not identical, since a female's breeding success nearly always jumped ahead with her first mating, whereas a male's sometimes didn't, if he copulated with an already-fertilized female).

One of Bateman's most influential findings was that beyond the initial mating, the reproductive success of males increased substantially with additional copulations, whereas that of females did not. Or rather, it increased much more slowly and reached the point of diminishing returns more rapidly. This is because a female's eggs, once successfully fertilized, cannot be fertilized any more, whereas a male can—at least in theory—fertilize additional females each time, until there are no more candidates.

But, in fact, females of many different species copulate many times, more than seems necessary to get their eggs fertilized. Most of the time, this apparent sexual excess takes place with the same male and within the identified social unit; that is, they are IPCs rather than EPCs. But as we shall see, EPCs may nonetheless be implicated.

Kestrels (sparrow hawks) copulate nearly 700 times per clutch. Female lions copulate on average every 15 minutes—day and night—throughout their frenetic, four-day estrous cycle. Generally, when such high mating frequencies are found, it is the female who is the initiator. Indian crested porcupines copulate every day throughout their estrus, pregnancy, and lactation, although in this species females are fertile for only slightly more than 1 percent of their estrus (and not at all when pregnant or lactating).

The traditional explanation for such behavior has long been that nonreproductive sex of this sort helps to maintain or enhance the pair-bond (as in at least one species, *Homo sapiens*). This seems eminently reasonable, but not sufficient. It is simple and, as such, worthy of distrust. Why does a high frequency of copulations enhance the pair-bond . . . assuming that it does so? On an immediate, "proximate" level, we might conclude that frequent matings generate sexual satisfaction and that this is reason enough. But why should females be wired to be "satisfied" in this way, especially if such behavior is not needed for fertilization?

One possibility—potentially disconnected from the question of EPCs— is mate assessment. Since it takes time and energy for a male to copulate, a female could assess a male's vigor and health by his sexual ability and incli- nation. Repeated matings also seem important in pair-formation. Newly- wed kittiwake gulls copulate more often than old married couples, a phe- nomenon to which nearly all human beings can also attest. Although male sexual inadequacy is occasionally cited by women as the reason for seeking a divorce, there are remarkably few animal parallels, that is, cases in which a female abandons her mate when he reveals himself to be sexually unen- thusiastic, inadequate, or just plain uninspiring. There are also few animal examples of what might be called the "Lady Chatterley option," after D. H. Lawrence's novel in which the wife of the paraplegic Lord Chatter- ley, erotically frustrated by her husband's incapacity, has a series of EPCs with her virile gamekeeper. Lady Chatterley is pictured as aroused and downright awe-struck by her lover's erect penis, although this may tell us more about Lawrence's erotic imaginings than about the realities of female sexual psychology.

On the other hand, something similar may be common in the natural word, motivated not so much by a mate's shortcomings as a lover as by his inadequacy as an inseminator: As we have seen, females of several species have been known to respond to their mate's diminished fertility by engaging in EPCs with other, more reproductively competent males. Sexual perform- ance as such could be used by females as a secondary sexual characteristic analogous to the "sexy son hypothesis": A sexually voracious female kes- trel, lion, or Indian porcupine may thus be more a tactician than a nympho- maniac, substituting her mate's sexual enthusiasm and capacity for what in other species is provided by an irresistibly dark throat patch, bright feath- ers, or impressively forked tail.

In the above cases, EPCs lurk in the background, an option for females if their males should fail their sexual exams. In others, EPCs may be more directly responsible. Females might well copulate repeatedly with a pre- ferred mate so as to swamp sperm already received—for example, via a forced copulation—from a less valued male. Even if not rape victims, females may occasionally acquiesce in EPCs because giving in is less costly than resisting (more on this later). In such cases, a female may then attempt to rectify her situation (i.e., inseminated by a less-desirable male) by repeat- edly copulating with a preferred one—who also happens to be her mate.

Or maybe females seek to ensure fertilization by a preferred male after having recently copulated with one or more other males whose genes they may esteem less but who, in return for being granted EPCs by the female, provide other benefits such as additional courtship feeding, the opportunity to forage on their territory, the promise of parental assistance for any young produced, and so on. High in-pair copulation frequency may also be a female's way of preventing polygyny or of keeping her male from partaking of his own EPCs.

Possibilities abound. For example, by copulating frequently, a female could increase her mate's confidence that he is the father, making him less likely to desert, more likely to help, and perhaps also less inclined to seek EPCs. Such enhanced confidence might or might not be justified: Frequent copulation probably does suggest a higher probability of paternity, but it isn't beyond the capacities of many animals to deceive their mates as to likely paternity in order to keep the male nearby and also more inclined to help take care of the kids.

Finally, an oft-copulating female might not simply be acting on her sexual motivation; she may also be *signaling* that she is sexually inclined. Throughout the animal world, such inclinations are generally at least somewhat correlated with ovulation and could therefore influence her mate to stick around and mate-guard . . . since EPCs loom as the alternative.

"It is a truth, universally acknowledged," writes Jane Austen in the famous opening sentence of *Pride and Prejudice,* "that a single man, in possession of a good fortune, must be in want of a wife." She goes on:

> However little known the feelings or views of such a man may be on his first entering a neighborhood, this truth is so well fixed in the minds of surrounding families that he is considered the rightful property of some one or other of their daughters.

Less universally acknowledged is the appeal of material resources to those daughters themselves, an appeal that may be so great that a man "in possession of a good fortune" may obtain copulations even if he is not single, or not "in want of a wife." In short, females may be inclined toward EPCs even if they are not prospecting for genes—if the material rewards are sufficient.

Red-billed gulls, studied off the coast of New Zealand, show a common pattern: Females who are well fed during courtship resist all EPC attempts, and they also remate with their partner the following year; on the other hand, females who had been poorly provisioned are especially likely to divorce in the future and are more likely to submit to EPCs—and even to solicit them. We have already reviewed a similar pattern among ospreys, another species in which males are expected to provide substantial calories and in which females engage in EPCs if their mates slack off.

It is at least possible that among species such as ospreys and red-billed gulls, which generally form long-term pair-bonds and which invest heavily in producing chicks and then caring for them, males are better off taking good care of their wives rather than spending their time and energy gallivanting. After all, in such species, males that switch partners end up with fewer successful offspring than do those who remain faithful and affiliated. In such cases, as red-billed gull expert J. A. Mills concludes, "the attentive prosper." Marital virtue—in this case, bringing home the bird equivalent of "the bacon"—may be its own reward, or rather, it can lead to higher reproductive rewards. Ditto for female fidelity, in all likelihood, but with the added proviso that if your male isn't a good provider, you might be better off mating with someone else.

There is a hard-hearted but deep-seated logic in females exchanging sex for resources and, if possible (or necessary), looking elsewhere for sex in the event that adequate resources aren't forthcoming from one's own mate. Among human beings, there is a long cross-cultural history of women being wooed by display of resources and being disaffected by material want. There is also a long cross-cultural history of men, whether married or not, using money to obtain sex from women, who also may or may not be married.

Moreover, even if prostitution isn't the world's oldest profession, it is without doubt one of the most widespread. And there is a difference only in degree between the prostitute who offers sexual services in return for a fee and the mistress who makes a similar exchange with a man who may be married but who has sufficient resources to provide her with various material benefits: an apartment, fancy clothing, jewelry, luxurious meals and vacations, and so forth. To be sure, the man "keeping" a mistress or visiting a prostitute is not attempting to reproduce, nor—in virtually all cases—is the woman. But in trading sex—often EPCs—for resources, people are almost certainly responding to an ancient connection, one that recent research has made increasingly evident.

The animal world is similarly filled with examples of females providing sexual access to themselves in exchange for resources controlled by males. Among the purple-throated Carib hummingbirds, males defend territories containing as many flowering trees of a particular, highly valued species as possible, aggressively chasing away other males. Interestingly, they also attack trespassing females . . . except those that solicit copulations. The technical article describing this system is aptly titled "Prostitution Behavior in a Tropical Hummingbird."

Another example comes from an odd little bird—the orange-rumped honeyguide—that lives in Nepal and loves to eat beeswax. Beehives, not surprisingly, are highly valued and are defended energetically by male honeyguides. To obtain the treasured beeswax, a female must first copulate with the proprietor male. In the world of orange-rumped honeyguides, "those that have, get," and those that don't, don't. Indeed, only hive-owning males do any mating, and one particularly "wealthy" male was observed to copulate 46 times with at least 18 different females during one breeding season.

In other cases, the concept of male as "provider" is taken even more literally, to the extent that the successful male may appear less enviable, at least to the human observer. Thus, certain spider species practice female sexual cannibalism, by which females profit directly from multiple copulations if only because they get to eat their numerous sexual partners. Although there is little obvious benefit from consuming just one male (especially since, in such species, the males tend to be much smaller than their voracious mates), multiple matings offer females the alluring prospect of a multicourse meal. It is conceivable that in such cases, the material benefit of munching on males—even more than mating with them—is what drives females to copulate as often as they do.

The insect world is especially rich in peculiar patterns of this sort, whereby females gain material benefits in unexpected ways. For example, in a species of moth, adults and eggs achieve protection from their predators via certain chemicals known as pyrrolizidine alkaloids (PAs). The moths cannot synthesize PAs; they have to obtain them from plants they eat or—more efficiently—from a male who has eaten these plants. Males secrete PAs into their spermatophores, large blobs of proteinaceous goo that are transferred to the female during mating and are consumed by her, whereupon they protect her and her eggs. Male moths attract females by use of specialized chemicals, pheromones, and the most effective pheromones—the ones most likely to turn on a female moth—are those containing large doses of PAs.

No one ever considered chimps monogamous. They have long been known, in fact, for the diverse, inscrutable, amorphous complexity of their sex lives. Alternatively, one might say that chimpanzee sexual arrangements were notorious for their simplicity: their *lack* of structure. But the former interpretation—complexity—is more likely accurate. Some tantalizing new discoveries about the sex lives of chimps have begun shedding light not only on *Pan troglodytes* themselves but also on yet another reason why females may engage in EPCs: recruiting lovers to provide enhanced care and protection of their offspring.

Female chimpanzees in heat will sometimes associate preferentially with one male, with the duo occasionally even isolating themselves from other troop members for several days to a few weeks. More often, a dominant male may monopolize (or attempt to monopolize) sexual access to a given female, especially when her anogenital swelling indicates that she is at peak fertility. Most commonly of all, females will copulate with many different males in their troop, although even in this case they seem especially prone to being inseminated by the most dominant adult male, since they are most likely to have sex with him when they are ovulating. At the same time, however, chimpanzees were at least thought to restrict their promiscuity (is "restricted promiscuity" an oxymoron?) to their own social group. No more.

It had long been assumed that the reason adolescent female chimps disperse from their social group is to avoid inbreeding. It may still be. But now it is also clear that females are not limited to the troop in which they reside; rather, they can—and do—mate with other males on the sly.

It is now apparent—once again, as a result of modern DNA technology—that females actively seek and obtain mating partners from outside their social unit. (Not quite an example of EPCs; we might call them EGCs—extra-group copulations.) It had long been known, from field observations, that females occasionally leave their troops for periods of a day or more, although no one knew why.

Recently, however, a study of chimps in the Tai forest, Côte d'Ivoire, found the answer. Females actively seek out mating partners from adjoining groups. In 13 cases, mother–infant pairs were analyzed for their DNA content and the results compared with DNA profiles obtained for in-group males. The results were startling: In 7 of these 13 cases, all the in-group males could be excluded as possible fathers! So, the baby chimps must have been fathered by males outside the mothers' group. Interestingly, all 7 of these females were known to have left their troops during their estrous period, precisely when the infants in question would have been conceived. It is also interesting that such absences are brief—in 4 cases, only one to two days— and, moreover, matings with nongroup males must be exceedingly furtive: During 17 years of continuous observations, the sharp-eyed researchers saw nary a one! (Without the DNA analysis, we would have had no way of knowing about this furtive aspect of the love lives of female chimps.)

The researchers suggest that this behavior by females allows them to choose from a wider variety of potential mating partners while still retaining the resources and social support of their in-group males. Another major possibility: They gain toleration of their young when different troops interact. Males may well say to themselves, in effect: "I remember this female, an old flame from several months ago. So maybe, just maybe, this cute little baby is my kid!"

Paternal perceptions of this sort, whether accurate or not, may turn out to be especially important for the survival of young chimps. An ethically troubling discovery—deriving initially from the pioneering research of primatologist Sarah Hrdy—has been that many animals practice infanticide. In brief, the pattern is as follows: Among polygynous species, when the harem-keeping male is eventually deposed, the newly ascendant male not uncommonly embarks on a grisly policy of killing the nursing infants. Although despicable by human moral standards, such behavior makes "good" evolutionary sense, since after their youngsters are eliminated, nursing mothers quickly resume ovulating, whereupon they are likely to mate with the new harem-keeper . . . despite the fact that he murdered their offspring. Insofar as the unfortunate infants were sired by the preceding male, their fate is of no biological concern to the newly ascendant infanticidal male. He is interested in his own progeny, not someone else's.

Female langur monkeys have even evolved an interesting counterstrategy. If a female langur is in the late stages of pregnancy when a male takeover occurs, she may undergo a "pseudo-estrus," developing swollen genitals and a sexual appetite for the new harem-keeper. Then, when her offspring is born, the adult male is more likely to act paternal than infanticidal.

Among many species, including chimpanzees and numerous other primates, the danger of infanticide is not limited to the aftermath of male takeovers. It is ever-present whenever two groups meet. However, as Sarah Hrdy has pointed out, given that even hard-hearted adult males are concerned about their own progeny, it may be that by copulating with more than one male, females introduce a degree of strategic uncertainty (or even erroneous confidence) as to whether a male who has enjoyed a female's sexual favors may accordingly have fathered her offspring. If so, then EPCs might serve as a kind of infanticide insurance, a means whereby females purchase a degree of immunity for their offspring.

Such a policy is not cheap. If purchased by chimpanzee females, it is at the cost of substantial risks. For example, females are at risk of predation, especially by leopards, while in transit between groups. Since male chimps are larger than females and physically dominant over them, females can be injured, even killed, when approaching a strange troop. In addition, there is potential risk if home-group members (especially males) discover that a female has been mating outside the social unit, although it isn't clear exactly how they would find out, nor has such a discovery ever been documented.

Bear in mind that in addition to their occasional trysts, female chimps also typically immigrate, as adolescents or occasionally as adults, into a troop different from the one in which they were born. It may be that, at such times, their prominent sexual swellings buy chimpanzee females a crucial degree of social acceptance. (It is an interesting fact that newly immigrant females often keep their sexual swellings for an unusually long time, and when they have recently changed troops, even pregnant females often produce swellings . . . something that does not normally happen when they are less at risk.) It seems likely that estrous swellings are important "safe-conduct passes" when an adult female encounters strange males.

It would seem that males from the home troop would be especially sensitive to the absence of females, particularly if such absence correlated with subsequent pregnancy. But at least in the Tai forest, females that mated with neighboring-group males were not absent from their home group any more than were those that mated exclusively within the troop, who occasionally depart for other reasons, including "legitimate" consortships with in-group males. So a female's absence from the group does not yield reliable cues as to whether she has been inseminated by outsiders. This may be quite important, since it is evident that, in other circumstances, chimps are capable of infanticide directed toward offspring obviously sired by outside males.

Hrdy's original suggestion was that females may engage in multiple mating so as to dupe males as to their paternity and thus obtain protection from infanticide. She has since expanded that notion, suggesting that females may also be inclined to mate with the male(s) best able and most likely to *protect* their young. From here, it is not much of a stretch to envision males displaying a degree of benevolent behavior toward a female's existing offspring as a part of their courtship tactics. It is well documented that among many primates living in multi-male troops, females have a mating preference for one particular male, or sometimes two.

University of Michigan primatologist Barbara Smuts has argued that female baboons mate preferentially with males with whom they have already established close "friendships" and that, in return, the females gain protection from aggression by other males toward themselves and their offspring. She reports that 91 percent of the time when a male defended a female or her young children from other baboons, he was the female's friend. Among the tiny rain-forest primates known as cotton-top tamarins, males carry their benevolence toward their sexual partner's existing offspring to great lengths: They often mate while literally holding the adult females' children.

Given that, among primates in particular, adult males are so often a threat to a female's offspring by a different male, it is not surprising that a male is more likely to win his lady-love's heart by showing that he is benevolently disposed toward her present offspring . . . or, at least, is unlikely to kill them.

This idea can be expanded further. It appears that females may often choose their EPC partners with an eye toward finding lovers who will not merely refrain from killing their offspring and will defend them from other bloody-minded males, but who will also contribute other forms of direct paternal care.

Female barn swallows, for instance, have an opportunity to assess the quality of a male's future parenting: by the type of nest he has built. In this species, nest-building occurs after mating. It appears that effort expended in nest-building serves as a "post-mating male sexual display," whereby males indicate to their females that they are ready and willing to invest in reproduction. It turns out that female barn swallows actually invest more in reproduction when their mates have constructed a large nest. Female barn swallows, you may recall, prefer males with long, deeply forked tails, even though such males provide less paternal care than do their less forked-tail counterparts. So, if you are a male barn swallow, genetically bequeathed a relatively short tail, all is not lost: To some extent, you can make up for it by demonstrating by your assiduous nest-building that you are still worth a female's attention. This is precisely what barn swallows do: Short-tailed males spend more time and effort nest-building than do long-tailed males.

Female appreciation of males that are potentially good parents is not limited, incidentally, to the comparatively brainy birds and mammals: Among fish known as the sand goby, females are known to reject dominant males in favor of those that are good fathers. Male sand gobies do all the parenting. On the other hand, in species with biparental care, a female who succeeds in deceiving her social mate about an EPC with an especially "paternal" male might receive a double dose of child-rearing assistance: from her social mate as well as from her sexual partner—that is, from her deceived "husband" as well as the actual father of her offspring.

We have already seen how in a small European songbird, the dunnock, males provide paternal assistance in proportion as they have copulated with the breeding female. Switching now to the female's perspective, it is entirely possible that a female dunnock doles out copulations to the males in her entourage, thereby leading each to believe that he is the likely father and to respond by behaving parentally—in particular, by providing food for her nestlings and, if need be, by defending them from predators. (This is reminiscent of the fabled grandmother who hugs her many grandchildren, whispering to each, in turn, "*You* are my favorite!")

It might therefore be in the interest of female dunnocks to mate with a large number of different males, so long as each can be similarly deceived. For their part, males, especially if socially dominant, attempt to monopolize the sexual attention of females, thereby maximizing their own reproductive success. As a result, female dunnocks engage in vast amounts of copulatory solicitation—sometimes more than 1,000 copulations per clutch. Not coincidentally, male dunnocks also have huge testes.

Another intriguing bit of dunnock lore: As part of precopulatory behavior, the male pecks at the female's cloaca; she responds by ejecting a small drop of semen, the remains of her most recent copulation (likely with a different male). Probably she does this to convince her current mating prospect that with his rival's sperm unceremoniously extruded, he has at least a chance of fertilizing some of her eggs.

Such deception is an option especially available to females that are fertilized internally, such as birds and mammals. In these cases, a male can never be certain that he has fathered a female's offspring: "Mommy's babies; Daddy's maybes." So, by permitting or even soliciting copulations, female birds or mammals may essentially dangle the prospect of paternity before gullible males. There are also many species that partake of external fertilization, as in numerous fishes and salamanders. In such cases, eggs and sperm are extruded into the surrounding water, which—despite the greater vulnerability of gametes to that water—at least gives males greater confidence of their paternity.

It is one thing for human fathers to be in the delivery room—and thus, perhaps, more bonded to their children as a result of this innovation—quite another for a male frog or fish to be literally present at the moment of conception, which happens in "public" instead of in the private confines of his mate's reproductive tract. At the same time, external fertilization can challenge females to ensure that their eggs get fertilized at all, especially when there is a strong water current that threatens to dilute their partner's sperm. One species of catfish has evolved a remarkable solution to this problem: Female *Corydoras aenus* practice a unique form of oral sex: sperm drinking, in which they place their mouth around the lower abdomen of the male. His sperm pass with extraordinary speed as well as digestive immunity through the female's stomach and intestines to emerge from her anus, whereupon they fertilize her eggs in a protected space created when the female curls her body and her pelvic fins.

Among fish and amphibians—in strong contrast to mammals—it is quite common for males to provide parental care. This makes sense when we consider that, in cases of external fertilization, male fish and amphibians commonly guard the eggs from the moment they are extruded from the female's body; thus, they are able to ensure that they are the fathers. Male birds, too, despite being internal fertilizers, often act paternally, although one especially revealing experiment showed how such tendencies are vulnerable to being exploited by savvy females, especially when such females are in need of help. Male pied flycatchers (another species of small songbird) were removed from the nest shortly after their females had laid eggs. These females—suddenly finding themselves potential single mothers—began soliciting copulations from other males, at least some of whom then helped the females rear broods, even though they were not the genetic fathers. Most likely, the males were duped; certainly, they are never observed to assist in the rearing of broods when they have *not* copulated with the mother. The pattern is reminiscent of those female langur monkeys, suddenly presented with a new and potentially infanticidal male, who respond with a seductive pseudo-estrus. Although female pied flycatchers do not have to worry about infanticide, they nonetheless respond to the loss of their "husbands" in a langur-like way: soliciting copulations from other males, some of whom are thereby recruited to assist with the child-rearing.

T he sad truth is that sometimes sex can be a chore . . . especially, it seems, for females, who must cope with males who are not only more ardent but often physically dangerous as well as demanding. It seems reasonable, therefore, that females may occasionally engage in EPCs not because doing so is in their interest, but because it is simply easier for them to go along than to resist. Such cases fall short of rape but are nonetheless examples of sexual coercion.

For example, if a female bird is sexually accosted while incubating, it is possible that resistance on her part might damage her eggs. Since most male birds lack a penis, they are probably unable to physically force a copulation. But they might nonetheless coerce an EPC by, in effect, threatening to destroy the victim's existing eggs unless she complies. Thus, among some species—including mallard ducks and indigo buntings—when females engage in EPCs but typically do not solicit them, extra-pair copulations may occur simply because the cost of just saying "no" is too great.

There may even be a physical cost to saying "yes." Among white ibis, marsh-dwelling birds with unusually graceful down-curved bills, EPCs are common—and so is theft. Male ibis never contribute nesting material to their extra-pair mates; instead, they often steal such material from them! A researcher of ibis mating behavior observed 164 cases of nesting-material theft. Of these, 82.5 percent occurred immediately after an EPC attempt, with the male intruder—evidently not content with trying to steal a copulation—also stealing from the nest.

Adding to the burden of EPC-ing females is violence, even when the female does not resist: 16 percent of all EPC attempts in this species involve males attacking females, jabbing and battering them with the tips of their bills, sometimes removing feathers and drawing blood. The researcher reports:

> Females that were chronically attacked, usually by particular males, would become wary and leave their nests at the approach of any males other than their mates. As a result, their nests were potentially more prone to predation . . . and nesting material theft than the nests of unabused females.

Interestingly, this behavior was not characteristic of EPC-obtaining males generally, but only of certain individuals: Fewer than one-quarter of all males accounted for more than one-half of all such attacks, suggesting that certain males were especially prone to "rough sex" as part of their EPC tactics; it is also noteworthy that females tended to avoid such males. At the same time, it is possible that if these males were not such aggressive brutes, they wouldn't obtain any EPCs at all.

Here is another example of females acquiescing to multiple matings, if not EPCs as such, in order to avoid the hassle of *not* mating with their importunate suitors. Courting male Sierra dome spiders come and squat in the web of nubile females, where they proceed to steal substantial numbers of prey. Only after females finally break down and mate with them do the freeloading males depart! Virtue may be its own reward, but for a beleaguered female Sierra dome spider, an even greater reward comes from giving the importunate males what they want . . . so they'll get out of her hair or, at least, her web.

At the same time, sometimes a little sexual jealousy can be a useful thing. Female eider ducks, for example, enjoy a higher feeding rate when they are attended by a guarding male. In these circumstances, the female is subjected to less harassment by other males. R. E. Ashcroft, the biologist who first described this phenomenon more than 25 years ago, believed that he had identified an additional function of the pair-bond—diminished harassment—in this species. With growing awareness of the importance of EPCs, it is now possible to see mate-guarding and even the pair-bond itself as a response to the risk of EPCs, with such possible benefits as improved foraging efficiency likely to be at most a secondary by-product of the primary reason for such behavior: efforts by males to thwart EPCs.

Similarly, the mating choices of females may be driven by the payoff that comes from keeping other importunate males at bay. We have already considered the hypothesis that female primates may be inclined to mate with more than one male because, by doing so, prospective mothers may recruit a cadre of expectant-father candidates who might therefore inhibit their otherwise infanticidal tendencies. Moreover, they might even choose males especially likely to provide protection for their offspring. Such cagey sexuality—assuming, of course, that it occurs at all—need not be limited to primates. For example, female mallard ducks paired with attractive males produce heavier chicks than when paired with unattractive males. This might conceivably be due to attractive males producing healthier offspring, but the actual reason appears to be even more straightforward: Females paired with desirable males lay larger eggs. No one knows the mechanism involved, but it may be as simple as that females mated with more socially dominant males aren't harassed as much by other males and are therefore able to feed more efficiently. (More than a few women are familiar with the tactic of choosing to spend the evening with the most physically imposing man at a bar or nightclub, thereby purchasing a kind of social immunity for themselves . . . at least insofar as being bothered by *other* males is concerned!)

Maybe one reason attractive males tend to invest less in their offspring is that their females tend to invest more. There is evidence from a number of different species that females often lay more eggs, each of which is larger and heavier, and that they also feed their chicks more actively when mated to an attractive male. Under these conditions—with females eagerly picking up the slack, or at least the pace—perhaps it isn't so surprising that males invest less! "Peahens Lay More Eggs," proclaimed the title of one noteworthy research article, "for Peacocks with Larger Trains."

A mong human beings, divorce is strongly correlated with EPCs, and vice versa. Most people assume, in fact, that EPCs, once discovered, lead to divorce. Indeed, adultery is among the most oft-cited grounds for marital dissolution.

But causation may often go the other way: Couples heading for a breakup are probably more likely to have EPCs. Carry this one step further—as some researchers have been doing in their studies of animals—and we get this: One reason for engaging in extra-pair copulations may well be to explore the possibility of divorcing one's current mate and reaffiliating with the EPC partner.

Sometimes, this might simply be another case of females in particular taking out a kind of insurance, analogous to the patterns of "fertilization insurance" described in Chapter 3. In this case, EPCs may be a suitable strategy for a female who might find herself divorced, or widowed, in the future.

Even for the linguistically squeamish, it should be no more incongruous to speak of animal "divorce" than to describe animal "courtship," "mating," "dominance hierarchies," "territories," "reproduction," "parental care," or, for that matter, to employ other words also used to describe the activities of human beings, such as "eating," "sleeping," "migrating," "digesting," "copulating," "defecating," "respiring," and "perspiring."

First, let's note that divorce is not unknown among animals; just as mateships are formed, they can be broken, and not only by the death of one of the partners. Some animals (most eagles, geese, beavers, and possibly foxes) are believed to mate "for life," but we are finding that even in these species, individuals occasionally desert their mate and establish a relationship with someone else. Take, for example, gibbons, those long-armed apes of Indonesia and Southeast Asia whose aerial acrobatics make them favorites at zoos and who have been admired for another reason, too: They were long considered paragons of lifetime monogamy. It is now clear, however, that permanent gibbon monogamy is a myth. Thus, one study has referred to their "dynamic pair-bonds," which is a nice way of saying that the "pair-bond" among gibbons turns out to be less a bond than a band ... a rubber band. Gibbon mating systems are so "dynamic" as to make us conclude that gibbons are swingers after all, and not just in matters of brachiation. Thus, of 11 heterosexual pairs observed in a long-term study in Sumatra, 5 were severed when one individual left its mate, typically to join a neighboring adult of the opposite sex.

There are many possible reasons for pairs—animal no less than human—to break up. These include incompatibility between them, either partner sensing that it has a better option, recognition that one or the other has made a mistake, a third-party intruder who forces a pair member out, and so forth. It used to be thought that animal divorce resulted simply from incompatibility (behavioral, genetic, physiological, even anatomical) between the partners, in which case a breakup would be beneficial to both. Thus, a renowned study of cliff-nesting gulls known as kittiwakes found that the best predictor of divorce in this species was a failure to breed successfully the previous year.

An alternative view, recently advanced, is the so-called better options hypothesis, which suggests that divorce results from a unilateral decision made by one member of a mated pair seeking to improve his or her situation. This leads to the possibility that EPCs and divorce may be closely related, if the former arc a means whereby a female determines whether to initiate the latter.

Thus, divorce and EPCs could be intimately connected as part of a "mate-sampling" process by which females use EPCs to assess the quality of potential mates prior to divorcing their current partner. A research paper titled "Why Does the Typically Monogamous Oystercatcher . . . Engage in Extra-Pair Copulations?" answered that question as follows: to identify a better mate. There is, however, an alternative possibility. Maybe EPCs occur particularly when divorce is *not* an option! Assuming that most, if not all, living things are inclined to "move up" if the option exists, an equally consistent strategy would be that if you cannot, and if you are stuck with a partner who is somehow inadequate—or, rather, less desirable than someone else who might be available—then try EPCs.

A growing body of evidence suggests, however, that females engaging in EPCs may be trying to pave the way for an eventual switch of partners, strengthening a prospective pair-bond with a future mate. Spotted sandpipers, for example, are likely to pair with individuals with whom they have previously engaged in EPCs. Such behavior has been described as a "mate-acquisition tactic," and it is employed particularly by females. Similarly, in another bird species, the ocean-going razorbills, EPCs are evidently used for mate appraisal.

As we have seen, females of many species are now known to intrude into the territory of extra-pair males, where they solicit copulations. They may even be increasing their chances of gaining more EPCs by choosing to settle down someplace where there are many male neighbors. It has also been suggested that the tendency of many species to breed in dense social groups ("colonies") may be due to the insistence of females upon mating with males whose nest-sites are near the territories of other males specifically because this gives these females the opportunity to mate with other males if they choose! (If so, males would have relatively little leverage; a male who refused to participate and insisted on maintaining his solitary splendor so as to be unthreatened by the prospect that his mate might copulate with a neighboring male might end up solitary indeed.)

When eastern bluebirds have already succeeded in breeding together successfully and are "old married couples," mated for at least the second time, they are significantly less likely to engage in EPCs than are "newlyweds" breeding together for the first time. Maybe female bluebirds are less likely to chance an EPC if they are already "happily married." In bird species as diverse as lesser scaup (a duck), barn swallows, and indigo buntings, females paired with younger males are more likely to stray sexually; those paired with older males are less likely to do so.

Generally, divorce rate is positively associated with EPC rate; in short, divorce and adultery are closely connected in animals, just as in human beings. It seems likely that females suffer possible mate desertion if their partner discovers that they have engaged in EPCs (more on this shortly). But despite many efforts to test such a connection, virtually no research has thus far confirmed it. A comparative study, looking at many different species, found a positive association between divorce and the frequency of EPCs in many different species; this is certainly suggestive, but it could be due to something else. Thus, if divorce is more likely when there is some inadequacy in either individual or incompatibility between them, then this inadequacy or incompatibility could give rise to EPCs and also—independently—to divorce, rather than divorce arising *because* the male detected his female's infidelity and therefore terminated the relationship.

In our own species, it seems likely that the connection between divorce and sexual infidelity works in several ways: To begin with, adultery is likely to lead to divorce; indeed, it has long been the most frequently cited reason for marital dissolution. But at the same time, adultery doesn't occur in a vacuum. An otherwise happily married individual doesn't suddenly find him- or herself being adulterous—in our more antiseptic terminology, engaging in one or more EPCs—without any antecedents, as one might suddenly be struck by lightning, pulverized by an errant meteor, or victimized by a flowerpot accidentally dropped by someone from a window 10 stories overhead. More likely, even if the adultery has not been carefully planned, it is predisposed by a degree of vulnerability within the married pair. When either husband or wife perceives that something is wrong (even if just a little bit wrong, and even if the existing discontent is below the level of conscious awareness), there may well be an increased openness to "prospecting" for an alternative partner. In this sense, impending divorce—or, at least, problems within the marriage—can also cause adultery.

L et's turn now to some reasons for *not* engaging in EPCs. Some likely ones are the most obvious. Maybe a female's in-pair male is simply more desirable than any extra-pair prospects. Maybe her male's mateguarding is just too effective to permit any extracurricular copulations. Or maybe female choice is prevented by males in other ways. Although there is no doubt that female choice is important, often overwhelmingly so, there are situations in which females simply don't have much choice, typically when males are substantially larger and more aggressive, as in elk, for example, or elephant seals. In such cases, when males may be several times the females in size, as well as sporting potentially lethal weaponry (antlers, tusks, etc.), females are likely to go along with the male who gathers them together into his harem.

The physically impressive and behaviorally intimidating bull elk almost certain owes his attributes to male–male competition: Large, ferocious males succeed in besting other males who are a bit smaller and less ferocious. In the process, they can also work their will upon the females . . . who, after all, generally have little reason to complain, since it is in their interests to mate with males whose traits have brought them such success. (Because their offspring, in turn, will be likely to have these traits and enjoy similar success.)

In other cases, female choice and male–male competition coexist uneasily. For example, the following experiment was conducted with captive brown trout. Female trout were exposed to two physically separated males who differed in the size of their adipose fin (long thought to be the key trait used by female trout to choose their mates). After the female revealed her choice—by preparing her nest closer to one of them, nearly always the one with the larger adipose fin—the two males were released. They proceeded to fight each other for the opportunity to spawn with the female in question. The result was that fewer than half the females ended up spawning with the male they had chosen. So, although mate choice is a real phenomenon for female brown trout, such a choice—at least in the laboratory—can be overwhelmed by male–male competition. It is interesting to note that successful males had higher levels of androgens, suggesting that while females were choosing based on a physical trait (size of the adipose fin), male success in competing with one another can be influenced by other, physiological factors, such as hormone levels.

Female EPCs can also be constrained by direct costs borne by the females. In Chapter 2, we considered possible costs to EPC-seeking males: being cuckolded by other males while you are away from home, being injured by your lover's mate, simply wasting time and energy if you are unsuccessful or if your lover isn't fertile. There is reason to think that the downsides of EPCs are more pronounced for females.

One possibility is sexually transmitted disease. The risk is notorious, especially for human beings in the age of AIDS. (Although, clearly, venereal diseases preceded AIDS, with gonnorhea and syphilis—and, more recently, herpes—being ancient scourges of those whose sexual adventuring transcended monogamy.) It seems likely that EPCs among animals, too, pose potential health risks, although it is remarkable how little we know about venereal diseases among free-living animals.

Males are generally rather bold when out cruising for possible extra-pair copulations, whereas, in nearly all species, females are furtive, often downright secretive. "Open marriages" are exceedingly rare in the natural world; when females engage in an EPC, it is pretty much only when their partners are absent. Indeed, such secretiveness also explains why EPCs were generally not observed by biologists studying the behavior of free-living animals, even after literally thousands of observation hours. If females are attempting to deceive their mates, they are likely to be even more successful with human observers, who are probably far more obtuse! (To this must be added an important point about the intersection of science and psychology: Even in so strictly a "reality-driven" enterprise as natural science, most of us tend to wear intellectual blinders, often failing to recognize something until we first have an explanation for it . . . or at least, an expectation of it. Believing is seeing.)

The fact that EPCs are nearly always hidden strongly suggests that male detection of such behavior is costly for the females. For example, in a study of red-winged blackbirds in the state of Washington, Elizabeth Gray found that fully 78 percent of all EPCs occurred when the female was away from the territory where she was nesting. EPCs took place either on the territory of the extra-pair male or off the marsh altogether. That is, in this species, extra-pair trysts are most likely to occur either at "his place" or at some neutral site. (Are there red-winged blackbird equivalents to cheap motels?)

Since animal copulations often require only a few seconds, they are sometimes surreptitiously and seamlessly squeezed into the crevices of an otherwise normal day. In contrast to male–female social units, which are normally formed only after the extensive give-and-take of lengthy and conspicuous courtship, EPCs are the animal equivalent of what author Erica Jong memorably dubbed a "zipless fuck," a sexual exchange so quick and transitory that the participants hardly even bother to unzip their clothes.

British behavioral ecologist Nick Davies tells the following anecdote: A pair of dunnocks had been feeding together, hopping peacefully toward a bush. Reaching it, the male proceeded to one side, the female to the other. Once out of the male's field of vision, the female instantly flew into the nearby undergrowth, where she copulated with a different male dunnock who had been hidden there. Immediately afterward, the female flew back to the bush, emerged around the side, and rejoined her mate, all the while acting as though nothing had happened.

Similar patterns are often observed among nonhuman primates, especially various species of macaque monkeys: The female copulates rather hurriedly behind a rock, tree, or bush, while her consort is temporarily distracted or otherwise unaware of what is transpiring. Immediately afterward, the extra-pair male will often cover his still-erect penis with a hand, as though to hide the evidence from the cuckolded male. All the while, Mrs. Macaque assumes a pose of almost comical nonchalance.

Why all the secrecy?

One possibility is that the outraged male will attack the female, punishing her physically for her infidelity. Surprisingly, perhaps, this has very rarely been documented. One exception is David's research on the "Male Response to Apparent Female Adultery in the Mountain Bluebird." In this study, he attached a model of a male mountain bluebird near two different nests occupied by mated pairs while the male was away foraging. Upon returning, each male was confronted with the appearance that his female had been sexually unfaithful. In each case his response was to attack the model quite vigorously; not only that, but in one instance the male bluebird actually attacked his mate as well, pulling out two primary flight feathers. She left and was replaced by another female, with whom the male subsequently reared a brood. (The researcher, duly chastened, and not wanting to continue in the role of Shakespeare's Iago, did not interfere additionally in the connubial bliss of this particular Othello. His Desdemona was presumably still alive, although residing elsewhere.)

In this case, the male attacked his mate immediately after attacking the model of an intruding male; his chastisement of his seemingly unfaithful—but actually innocent—mate may have been due to unappeased aggression, which was aroused by the model but not fully discharged, since the model did not fight back. In the case of real-life EPCs, the outraged "husband" may be more likely to vent his fury on the other trespassing male than on the "errant wife." Also, each species presents a slightly different situation. Mountain bluebirds nest in empty holes in trees; because such nest-sites are rare, there is typically a reservoir of unmated female mountain bluebirds available to mate with a male who has driven out a suspected adulteress. Among most other species, males may not have the luxury of responding so aggressively. Their most effective countermeasure appears to be a withdrawal of parental assistance.

It is not a trivial threat.


Being a single parent is difficult, whether among animals or human beings. Indeed, the advantage of two parents over one seems to be the main reason why social monogamy occurs at all. Given that, in most species, females make the bulk of the parental investment, they are especially vulnerable to being left holding the bag—for our purposes, holding the babies—without assistance from their mate. In other words, among birds and mammals there is a much greater risk that females will be abandoned by their mates and left to be single mothers than that males will be stuck with being single fathers. And among the risk factors that seem likely to contribute to abandonment, EPCs rank very high. After all, males are unlikely to invest time and energy, or to undergo substantial risk, in caring for someone else's kids.

In Rodgers and Hammerstein's classic Broadway musical *Oklahoma!*, there is a marvelous duet sung by the two comic leads, Ado Annie (the "Girl Who Cain't Say No") and her betrothed, Will Parker. Will chastises Annie for her wild ways, demanding that, for them to marry, she must agree to be a chaste, submissive, dutiful wife: "If you cain't give me all, give me nothin', and nothin's what you'll git from me." Annie responds:

> I ain't gonna fuss, ain't gonna frown,
> Have your fun, go out on the town,
> Stay out late and don't come home till three.
> And go right off to sleep if you're sleepy . . .
> No use waitin' up for me!

Rather than exceptions, the Ado Annies of the world are abundant, and maybe even the rule. (Hence, this book). At the same time, it is noteworthy that later in the same song, when Annie asks, "Supposing that we should have a third one?" Will responds immediately: "He'd better look a lot like me!"

Female barn swallows who "cain't"—or, at least, don't—say "no," get less help from their mates: When females copulate repeatedly with other males, their own mates provide less assistance in rearing their young than when females rarely if ever copulate outside the mateship. When males reduce paternal care in response to female EPCs, females often increase their own maternal care, but frequently not enough to compensate entirely; as a result, their offspring are often short-changed. In other cases, compensation by the female is apparently complete, although females seem likely to suffer in the long run, probably through reduced survival rates during the non-breeding season or a diminished life span.

Although eastern bluebirds are normally monogamous—at least socially—when males were experimentally removed, the remaining single-parent females were as successful at rearing offspring as were dual-parent bluebird families. This ability and inclination on the part of female bluebirds to make up for male absence may have the paradoxical effect of encouraging males to wander in search of EPCs. It also suggests that even in some socially monogamous species, females might not be as utterly dependent on males as previous researchers had assumed . . . although, once again, it remains possible—even likely—that when females are forced to make up for their mate's absence, they ultimately suffer in the long run, perhaps by having a shorter life span.

The risk that males may reduce their parental contribution may limit the degree to which females pursue EPCs, especially in species with biparental care. Fulmars and kittiwakes are colonial-breeding seabirds for whom cooperation between males and females is essential in order for nestlings to be reared. Among these species, EPCs are notably rare, perhaps because the cost of detection is too high to be worth the risk.

On the other hand, the well-studied red-winged blackbirds are typically polygynous, but not necessarily monandrous. Since most females are mated to males who have other "wives," they are already adapted to caring for their young single-handedly. And not surprisingly, EPCs are comparatively frequent in this species, since females have less to lose. This is not to say that the male's contribution is trivial, however. In a population of red-winged blackbirds studied in eastern Ontario, nest defense is the major contribution made by fathers. (Blackbird chicks make a tasty meal for a variety of predators, including crows, gulls, hawks, weasels, raccoons, and mink). Previous research has shown that male swallows are less inclined to defend their young from a stuffed model predator (presented by the experimenter) when their females had earlier engaged in EPCs.

A five-year study of red-wings in eastern Ontario found that the greater the proportion of nestlings sired via EPCs, the less vigorous is the nest defense provided by resident males. Nests in which the putative father had fathered all the young had the highest success rate (number of fledging young). Those in which all the young were sired via EPCs had the lowest. Mixed broods were intermediate. Not only are resident males zealous in defending juveniles in proportion as they have fathered some of them, there is even some evidence that males who sire young via EPCs in other territories are more likely to defend those young.

These results have not been found in all cases. Indeed, even other populations of the same species, red-winged blackbirds, fall out differently. A study conducted by University of Kentucky biologist David Westneat, for example, found no difference in fledging success between nests with and without extra-pair young. In fact, Westneat found that broods sired by multiple males were *more* successful. Having found no difference in the feeding of nestlings in the two cases, Westneat concluded that mixed-paternity nests did better because the additional males provided, on balance, more paternal defense, not less.

These studies are not necessarily contradictory, however: Let's assume that out-of-pair male red-wings are more likely to help defend their potential offspring if they have had an EPC with the mother. Then it simply becomes a matter of whether the assistance that is forgone (from the resident male, whose confidence in parentage is diminished if he knows of his mate's EPC) is compensated by the additional defense that is obtained (from the extra-pair male).

In most cases, however, it appears that when a female engages in one or more EPCs, she runs a risk that her offspring will receive less paternal care as a result. The salient finding of one research effort was conveyed in the title of the article: "Paternal Investment Inversely Related to Degree of Extra-Pair Paternity in the Reed Bunting." In some cases, males evidently refuse to pay child support if their mate has copulated with someone else. After an EPC that took place in the middle of the female's fertile period, a male magpie abandoned the clutch that he and his mate had been incubating; shortly afterward, he initiated another breeding attempt . . . interestingly, with the same female. (A second marriage, it has been said, represents the triumph of human hope over experience; maybe something similar can be said of renesting, at least among magpies.)

Young male purple martins, which are often cuckolded by older males, are rather lethargic about feeding "their" nestlings, being significantly less attentive than the females. On the other hand, older male purple martins, cuckolded only rarely, are comparable to females when it comes to fulfilling their parental duties. (It is also possible, of course, that young males just aren't "into" feeding nestlings, whether they have been cuckolded or not.)

Behavioral ecologist Bert Kempenaers relates this story: After a male blue tit was injured—apparently by a sparrow hawk—his mate visited males in several nearby territories and was seen to copulate with at least one of them. She then laid a number of eggs, and a few days later, her mate (the injured male) died. Of the six nestlings produced, DNA analysis revealed that one had been fathered by the deceased male, three had been fathered by the extra-pair male seen to have copulated with the female, and the remaining two had been fathered by another male. Neither of the extra-pair males helped care for his bastard children, although it is interesting that one of them threatened a human observer visiting the widowed female's nest; male blue tits have never been reported to behave in this way toward a nest not their own ... so perhaps this male was displaying at least a modicum of parental inclination.

Aside from the potential downside that a cuckolded male may provide less parental care, EPCs may subject offspring to enhanced risk in another way: Without EPCs, all the nestlings would be fathered by the same male (as well as conceived by the same female, of course). As a result, they would all be full siblings, with a high probability of shared genes and, thus, shared interests in one another's success. With EPCs, some of the nestlings would have different fathers and would therefore be half-siblings, as a result of which their genetic relatedness is diminished by a factor of 2. It is increasingly clear that shared genes correspond to greater altruism and less selfishness; as a result, EPC-produced offspring, being less closely related, may well be less benevolently inclined toward one another. This, in turn, might result in less food sharing, less mutual defense, and generally diminished success. And it has already been found that when nestmates are less closely related, they tend to beg more loudly and selfishly; this, in turn, could attract additional predators.

It might also be costly for females to incur the jealousy of their mates, aside from the possibility of physical punishment or abandonment. Even being guarded by one's mate could have a negative impact, and it is not unreasonable to suppose that females who are prone to EPCs are liable to being guarded more closely. For example, male dunnocks shadow their female partners while the latter are foraging, and this results in a diminished food intake. Among dunnocks that are truly monogamous, as opposed to those groups consisting of two or more males and one female, males guard their females less closely, and so the latter get to forage unencumbered and thus more efficiently.

I t is time to end this chapter. But first, let's look briefly at a strange aspect of extra-pair reproduction on the part of some female birds. It goes by the indelicate term *egg dumping*.

Normally, we think of EPCs as producing a situation in which the male is cuckolded because "his" female has mated with someone else. As a result, some of the offspring produced are not his ... although they are still hers. But not necessarily. It occasionally happens that females deposit fertilized eggs in someone else's nest, leaving the foster parents to perform the child care. These cases are well known for a number of animals known as "nest parasites," species such as the European cuckoo or the North American cowbird, which typically do not care for their own offspring; rather, they sneak their eggs into the nests of unwitting "host" species.

This is not considered egg dumping; rather, it is a species-wide reproductive strategy, albeit a parasitic one. True egg dumpers are found in species that normally take care of their own offspring but in which a small number of individuals have opted for this tactic of semi-parasitism. Not surprisingly, female birds are more likely than mammals to foist their offspring onto another female. And among birds, a successful "dumper" can "cuckold" another female, no less than her mate. At the same time, it is also possible for the dumper to have copulated (via an EPC) with the mate of the victimized female . . . in which case, we might expect that this male would be especially solicitous of the offspring, assuming that he somehow knows that the eggs were deposited by his clandestine lover! There already exists a phrase in the technical literature that describes this situation in which the attending male, but not the female, is the genetic parent: *quasi-parasitism.*

We have long known that among many species of birds, there exists a kind of female underworld, a population of so-called floaters—nonbreeding individuals who are kept in their spinster status by a shortage of resources, usually suitable nest-sites. A recent discovery has been that these floaters are not quite so nonbreeding as had been thought. For example, one study of European starlings captured floaters by attracting them to artificially constructed nestboxes. Nearly 50 percent of these homeless, unmated, and ostensibly nonbreeding females were found to have either laid an egg in a nest or carried a fully developed egg in their reproductive tracts, indicating that they had mated with someone and were about to deposit their egg somewhere. Since they did not have their own nests, that "somewhere" was going to be someone else's nest! As a rule, floating females are significantly younger and smaller than normal breeders.

Generally, it isn't clear whether egg dumping is largely the work of floaters, but it quite likely is. Among the waterbirds known as coots, for example, this is well established. In any event, the likelihood is that it isn't only males that are prone to a "mixed reproductive strategy." Whereas for males such a strategy typically involves monogamy combined with extra-pair copulations, for females it includes the additional option of being a floater who dumps her eggs into someone else's nest (relying, therefore, on others to do the parenting). As a rule, however, egg dumping is unlikely to be a preferred strategy. Rather, in most cases it is probably an attempt by disadvantaged females to make the best of their bad situation, forced on them by the animal equivalent of poverty.

On the other hand, in at least one documented case, egg dumpers are the cream of the crop: Among goldeneye ducks, the oldest and strongest females not only maintain their own nests, they also lay additional eggs in the nests of other females. It appears that younger, weaker females are incapable of producing so many eggs; hence, they are less likely to dump and more likely to be dumped upon.

Although it is relatively uncommon, egg dumping nonetheless occurs in perching birds, such as starlings and house and savanna sparrows, as well as in ducks. One study found that among cliff swallows (which nest monogamously, but in colonies consisting of dozens, sometimes hundreds, of individuals), nearly one-quarter of all nests contained eggs laid by one or more females who were not incubating or feeding the young. The researcher concluded that egg dumping is a major downside of colonial life, at least among birds. Among mammals, females have complete confidence of their maternity (a certainty that males can never attain). But among birds, even females cannot know for certain that their nestlings are really theirs!

We can expect that prospective victims—both male and female—would be on guard to prevent such dumping . . . except, as noted, if the male had recently copulated with the dumper. Then he and she could conceivably be in cahoots. (In a novella by Joseph Conrad, a man's determination to adopt an orphaned child leads his wife to suspect—wrongly, it turns out—that her husband had fathered the child via an extramarital affair.) In Dr. Seuss's children's book *Horton Hatches the Egg*, Mayzie bird—"lazy bird"—leaves an egg in the care of our hero, a big-hearted elephant, who steadfastly guards it through thick and thin. Mayzie eventually returns and demands her egg, just before it hatches; righteousness is ultimately rewarded, however, since the creature that emerges bears a striking resemblance to generous, gentle Horton!

Maybe, while Horton the elephant was incubating Mayzie's egg, some of Horton's genes somehow leaked through the shell (a one-of-a-kind event). Or maybe Horton and Mayzie had earlier partaken of an EPC or two on the side (equally improbable, since mammals and birds do not normally interbreed, not even a little bit). In any event, Dr. Seuss's poetic justice satisfies our notion of what is fair: Mayzie the egg dumper gets trumped.

CHAPTER FIVE

# Why Does Monogamy Occur At All?

The eminent eighteenth-century English essayist and critic Dr. Samuel Johnson once wrote this about a dog walking on its hind legs: "It is not done well; but you are surprised to find it done at all." We might say the same about monogamy: Considering how many strikes there are against it, how wobbly most living things are when they seek to balance on monogamous limbs, it is remarkable that it is done at all. Since both males and females have reasons to deviate from it, why does monogamy ever occur?

To be sure, monogamy is rare. But it does happen sometimes, and so logic suggests that it must—at least on occasion—have something going for it. There are several possibilities, some of them unique to human beings and others that seem to apply to animals generally. Let's look at the general patterns first.

Better the devil you know than the one you don't. Maybe some animals (and, occasionally, people) form monogamous unions because they are, in a sense, conservative. After all, courtship and mating are risky, requiring that both partners venture out of their personal shell and become vulnerable to rejection, injury, bad choices, or just plain wasting of time and energy. Having done so once—and succeeded in obtaining a mate—it is possible that certain individuals might just elect to stop such anxious and risky prospecting and settle down to a life of cozy, comfortable domesticity. "If it ain't broke, don't fix it." Leave well enough alone.

On a more positive note, having found a reliable and mutually gratifying relationship, why rock the boat? More positively yet, it has been documented among animals that the longer pairs are together, the more likely they are to be successful at rearing offspring. This may be because experience and familiarity with each other make for better and more efficient parenting. Familiarity doesn't have to breed contempt; it can also breed competence (particularly competence at breeding!). On the other hand, causation may also work the other way: mates that remain together in succeeding years are generally more likely to be those that are reproductively successful. Thus, competence breeds contentment. And, in turn, commitment. Divorce, as we have seen, is common in nature, often correlating with whether the pair had been reproductively unsuccessful in the past. Among kittiwake gulls, for example, couples are more likely to split up after a year in which they failed to breed.

It is also likely that many living things—human beings included—have a sense of their own self-worth as measured by their potential success in what might be called the "mating market." One of the best-documented tendencies in animal (and human) courtship and mate selection is "assortative mating," which refers to the phenomenon whereby mates tend to be similar. To be sure, opposites attract—at least opposite sexes—but otherwise, it is remarkable to what extent individuals choose members of the opposite sex who are *similar* to themselves. Whether in matters of physical size, cultural background, intelligence, political inclinations, or overall degree of personal attractiveness, people gravitate toward mates who are like themselves. (How often have you met an unattractive woman married to a very handsome man, or vice versa?) A similar pattern is found in animals, too.

What does this have to do with monogamy? Just this: When a mated pair consists of individuals who are not only mated but also matched, there is probably a greater chance that their monogamy will persist. The greatest stability would presumably arise from a situation in which each partner actually is—or perceives itself to be—just a bit less desirable than the other! In this case, each would likely think that he or she has gotten a good deal (that is, a somewhat "better" mate than entitled) and would be unlikely to risk rocking the boat by reaching higher yet. The greater the disparity, the greater the chance that the more highly valued individual will attempt either to terminate the relationship or, failing that, to go outside the monogamous union.

Biologists have long known that females are generally the choosy sex and that males tend to be much less fussy. There are interesting exceptions, however. For example, when males perceive themselves to be especially desirable, they tend to become proportionally more demanding of high-class partners only. Not surprisingly, the same applies to females to an even greater degree, since they have something—their large eggs or, in the case of mammals, the promise of nourishing offspring during pregnancy and lactation—that males want.

Sometimes, however, the options for either sex are restricted simply by the force of circumstance. The result is a higher probability of monogamy, simply because there is little alternative. If there aren't many potential partners from whom to choose, or if it is literally difficult to get near anyone else, the likely result is a higher degree of fidelity—not because of choice or because the partners are especially virtuous, but simply out of necessity.

Some studies dealing with fish reveal a pattern of exceptions to female choosiness (or rather, examples of fine-tuning). In one case, researchers designed an experimental setup in which female fish had to swim against a current of water in order to get to different males. In this situation, males that had been unacceptable suddenly became highly attractive, *if* the normally preferred alternatives were unattainable.

In an earlier publication, the same researchers had reported that female fish use male coloration in mate choice, thereby avoiding parasitized males (who are unable to produce bright coloration). These findings led Silvia Lopez, then a doctoral student at Oxford University, to ask the following question: What about the effect, on *females*, of being parasitized? Thus, what happens when the females themselves are less healthy and, thus, less desirable? Do they become less choosy? For her research, Lopez chose guppies, the common, brightly colored aquarium fish.

Guppy males engage in one of two courtship strategies: attempting to persuade females to mate with them by vibrating and bending their colorful bodies (traditional guppy courtship) or attempting to sneak a copulation. With the first strategy, brightly colored males have a definite advantage, being preferred by the females. This seems doubly beneficial for the females, since they not only increase their chance of producing young that are brightly colored because of inheriting their father's sexy genes, but they also avoid parasitized males, who might infect them or their offspring and whose parasitized state may indicate a genetic weakness when it comes to keeping parasites at bay.

Lopez noted, however, that virtually nothing was known about the effect of being parasitized on the behavior of females. So she established populations of virgin guppies, some of whom were parasite-free, and some, infected. Her findings? They are indicated by the title of her research report: "Parasitized Female Guppies Do Not Prefer Showy Males." Beggars, it seems, can't be choosers ... even among guppies. Maybe parasitized females lack the energy to assess several different males and then choose the best one. Or maybe these females recognize that they are, in a sense, "damaged goods" and lower their sights accordingly. It is even possible that their behavior is somehow manipulated by their parasites: Bear in mind that parasites have an interest in getting themselves transmitted to new hosts, and unparasitized males may well be somewhat resistant to them, so the guppy parasites may opt for parasitized males as a more vulnerable target. The biological world has seen other outcomes that are at least as devious.

In any event, the result in this case is that females, no less than males, may be limited in their choice of mating partners by factors related to their own desirability. It probably wouldn't be wise to anticipate a glowingly faithful monogamous future for a healthy female mated to a heavily parasitized male or vice versa. But matched pairs—in which male and female are both healthy or both unhealthy—may be destined for (or doomed to) a lifetime of monogamous bliss.

William James is said to have composed this ditty, although with insights inspired by opium rather than evolutionary biology:

Higamous hogamous, woman monogamous
Hogamous higamous, man is polygamous.

And in her "General Review of the Sex Situation," Dorothy Parker put it this way:

Woman wants monogamy;
Man delights in novelty.
Love is woman's moon and sun;
Man has other forms of fun.
Woman lives but in her lord;
Count to ten, and man is bored.
With this the gist and sum of it,
What earthly good can come of it?

We now understand that women (females) are not all that monogamous, and that men (males) aren't always polygamous. But we also know that males—because they are sperm-makers rather than egg-makers—are more likely, in general, to seek multiple mating opportunities than are females. It is also generally true that the biological success of a male is more likely to be diminished by his female's EPCs than her success is by EPCs on his part. (This is because males are liable to being cuckolded—reproductively excluded—by their female's extra-pair matings, whereas a female will continue to be the mother of her offspring even if her male copulates with one or more additional females.) Nonetheless, a male's EPCs can still have consequences for his social mate, and nearly always these consequences are negative. The male may be injured during his gallivanting and thus less able to help with his domestic responsibilities. He may contract a sexually transmitted disease and then infect his mate. He may elect to leave his mate, having discovered a more desirable partner. And—probably the greatest risk, because the most likely—he might find himself devoting time and effort to his lover's offspring, providing benefits that are deducted from his "official" mate and her offspring.

This risk is greater yet in species that are sometimes monogamous, sometimes polygynous—that is, in which males occasionally succeed in transforming an EPC conquest into an additional, full-time reproductive partner. We can therefore expect that in such cases, females will be strongly motivated to keep their males from engaging in EPCs—and even more strongly inclined to prevent their switching from monogamy to polygyny. The end result? Males may seek multiple mates and even polygyny but end up with monogamy because of the intervention of their females, who won't let them bring their girlfriends home.

In a species of lizards common in the Southwest, females defend small territories from which they exclude other females. Males cannot monopolize more than one female because females are so antagonistic to one another that their territories are dispersed.

In most such cases, it appears that females are moved to keep other females at bay when the males have something of value—usually, parental care—to contribute. But sometimes, even when males provide virtually nothing (besides sperm) to their offspring, monogamy remains the most frequent system; for example, among willow ptarmigans, grouse-like residents of mountain and arctic habitats, it seems likely that females are too dispersed to permit bigamy. These animals may also be too aggressive toward each other to accommodate a second female within their domain . . . much as the male might want to.

It isn't clear what the "married" female ptarmigan would lose by permitting her husband to have a second wife, since his contribution is essentially nil; probably it isn't a matter of having to share *him,* but, rather, having to share limited food resources with another female and her brood. Similarly, among eastern bluebirds (normally monogamous), males provide little more than a nesting cavity. What generates social monogamy in the bluebird world is the fact that suitable nest-sites are few and far between, and—acting in their own self-interest—female bluebirds won't share them. (It must be pointed out, however, that polygynous females do not always suffer an obvious cost when their male takes another mate; it is possible, that in such cases, males are able to add one or more additional females to their harems precisely because doing so imposes no extra cost to existing females. When this is true, the current wives do not seek to interfere.)

In the world of European starlings, breeding occurs in socially monogamous pairs, or in bigamous trios consisting of one male and two females, or occasionally even trigamous quartets of one male and three females. The latter arrangements benefit males but are disadvantageous to females, since multiple-mated males provide less parental care per nest than do their monogamous counterparts. In an experiment, males were given the opportunity to form bigamous or trigamous relationships, or to remain monogamously mated to their current female. Males who remained monogamous when they apparently had the opportunity to attract a second or third female were those mated to unusually aggressive females. (This aggression, incidentally, was directed toward the potentially home-wrecking females, not toward the male.) In starlings, at least, female aggression is a reliable predictor of male mating status. So, if you are a male starling with a yen for multiple mates, better choose them from the mild-mannered end of the female spectrum.

In another bird species whose breeding system is all over the marital map, the European dunnock, it appears that female song serves to deter rival females. Just as males sing to attract females and post a vocal "no trespassing" sign to other males, females evidently can inform other females that "this male is taken" . . . and not only that, he is taken by a feisty female.

Such feistiness can become lethal. Spanish biologist José Veiga studied house sparrows breeding monogamously, yet close to other mated pairs. He wanted to know why 90 percent of them were monogamous. Veiga could eliminate several possible explanations: House sparrow monogamy was not due to the fact that an attentive male is needed for rearing young (bigamously mated females reared as many offspring). Nor was it due to female reluctance to mate with already-mated males. And it wasn't because there weren't enough unmated, available females. Why then? Because mated females were aggressive toward other females: Moving nestboxes closer together induced females to attack each other. Earlier, Veiga had found that a female house sparrow will sometimes kill the young of another female who is sharing her male's sexual attentions . . . shades of the movie *Fatal Attraction*. Given that female house sparrows are willing to carry their sexual rivalry to such extremes, there is evidently a certain proactive, nonviolent wisdom in a healthy degree of female–female repulsion.

A growing number of studies have in fact confirmed that among animals, females are frequently aggressive toward potential "home-wreckers." In Chapter 4, we examined the peculiar phenomenon of egg dumping, whereby females occasionally deposit eggs in another's nest; female–female aggression among birds may often be motivated by anti-dumping watchfulness on the part of mated females. Thus, when a female bird chases other females away, she might well be less concerned that the intruders will attempt to mate with "her" male than that they have already mated with someone and are now seeking—like Horton's nemesis, the infamous Mayzie bird—to dump the offspring in her lap (or rather, nest).

Whatever its origin, it is likely that such female–female alertness gives a boost to monogamy as well, simply by making it difficult for a male to affiliate with more than one female. One might also recall those animals—birds, in particular—for which EPCs often precede divorce. By extension, therefore, it could pay a female to break up any such liaisons, so as to make it less likely that her mate will desert her and set up housekeeping with a new paramour. (It is pretty much taken for granted that male–male aggressive alertness serves in many species to keep social monogamy from becoming polyandry, insofar as, by mate-guarding, males keep "their" females from affiliating—if not EPC-ing—with other males.)

Among mammals, monogamy also seems to be maintained, on occasion, by female–female aggression: There is some tendency among monogamous mammals for the females to be aggressive, especially toward other females. The feistiness of resident female beavers, for example, seems to keep other females away. If Madame Beavery were less aggressive, perhaps the species would be polygynous.

Female vigilance of this sort is evidently common in nonhuman primates, too. Primatologist Barbara Smuts reports that female baboons behave aggressively toward other females that show sexual interest in their consort partner or in which their partner shows sexual interest. There are also many cases in which female mammals—especially among the social canids, including wolves, jackals, and African wild dogs—prevent subordinate females from breeding. A wolf pack, for example, will typically contain only one breeding female, who also, not surprisingly, is socially dominant over the other females. Only if this alpha female is removed will the others copulate and produce pups. As a result, the dominant male of the pack may be *socially* polygynous (that is, he "has" a harem consisting of more than one female) but be kept *reproductively* monogamous by the breeding female's impact on her female rivals.

We have been discussing cases in which monogamy is maintained by female aggression toward other females, thwarting the polygynous inclinations of their mates. It seems only logical that there should also be species in which females impose monogamy by aggressive behavior toward their males when the latter show a hankering for EPCs or—worse yet, from the female's perspective—one or more additional mates. We are only aware of one such case, the burying beetle *Nicrophorus defodiens*. In this monogamous insect, male and female cooperate in the unseemly task of burying a dead animal—

commonly a mouse—upon which the female will lay her eggs. Having initiated the burying of a carcass, which the female has duly anointed with her fertilized eggs, male beetles are sometimes moved to emit mate attraction chemicals (pheromones). If successful in this maneuver, the male will have lured a second female, who, after copulating with him, would add her eggs to the appealing lump of moldy mouse meat already occupied by the initial female's developing brood. As a result, the freshly hatched maggots of female number two would compete for food with the offspring of female number one, who, after all, got there first, has already "shot her wad" of eggs, and, moreover, has also expended effort in getting the dead mouse prepped to become grub for her own grubs.

So what does the "outraged" female beetle do? As soon as she detects that her mate is beginning to feel frisky and inclined to broadcast his pheromones, she rushes over to him, pushes him off the perch from which he is attempting to broadcast his love potion, and often bites him as well! Not surprisingly, this interferes with his ability to fulfill his polygynous hopes. To demonstrate this, researchers tried tethering number-one females so that they could not reach their pheromone-secreting mates. Thus liberated from their jealous wives, male burying beetles joyously broadcast their sexy scents for prolonged periods—and were rewarded with additional girlfriends.

These more or less direct examples of female–female competition should not blind us to the existence of subtler forms of female–female sexual maneuvering, all of which are also potentially implicated in the maintenance of monogamy.

Thus, jealous females are not limited to outright aggression in their quest to keep their male partners monogamous. Happily married starlings were experimentally presented with an extra nest-site at different distances from their current nest. Among these birds, as among many others, nest-sites are in short supply, so the prospect of two nest-sites is pretty much equivalent to the prospect of exchanging monogamy for bigamy. But when such additional mating opportunities were close to their existing domicile, very few male starlings became polygynous; the farther the new potential love-nests, the more likely that they were employed. In fact, most males became polygynous when there was enough distance between their new lovers and their preexisting mates—but only then. This suggests that what constrains mated males to monogamy is the existence of their present mates, but, by itself, it doesn't suggest how a starling wife interferes with her mate's polygynous designs.

There are several ways for a female to interfere with her mate's efforts at acquiring additional females. One of the most interesting involves her own sexuality: A female starling is especially likely to solicit copulations from her mate when he is actively courting other females! There is nothing as likely to make a female starling feel friendly and sexy than the prospect that her mate is showing interest in another female. Equally interesting: A large proportion of these solicitations are refused by her partner, but they nonetheless succeed in causing any prospecting females to depart.

Not surprisingly, however, the male is often less than entranced at such a show of sexual solicitude from his mate. Belgian biologists Marcel Eens and Rianne Pinxten observed 14 instances of male starlings attacking or chasing their females after the ladies had solicited a copulation. Some of these chases were quite vigorous, lasting more than a full minute. In most of them, males had previously landed on another nestbox and were singing, evidently attempting to attract an additional female. By contrast, of ten males who were monogamous and showed every sign of remaining that way, not one was ever observed to attack his female when she solicited a copulation. Why do would-be polygynous males often refuse to copulate? Two likely reasons: First, by doing so, they give away to any potential new mate the fact that they are already mated. Secondly, by copulating with an existing mate, they might so deplete their sperm resources that they would be less likely to fertilize any newcomer.

(It would be interesting to know if anything comparable were true of harems in that peculiar species *Homo sapiens*. It is widely acknowledged that women in human harems competed with one another for resources and benefits, especially those potentially useful to their children. But did they also compete sexually for the erotic attentions of the harem-keeper? If so, then perhaps we should pity the poor, sexually beleaguered sultans, who were likely to have been doubly exhausted, not only because of the simple, numerical demands mandated by their various wives and concubines, but also because of the extra pressure generated if each female—because of her competitive relationship with her co-wives—was especially desirous of his sexual attention.)

Why don't female starlings simply attack would-be rivals? Probably for the same reason that human harem-members are unlikely to do so: It is bad politics, likely to evoke the wrath of the male. In such cases, subtle, seductive tactics are probably more successful. Thus, mate-guarding—of the sort we saw in Chapter 2—is not a males-only tactic. Females sometimes go out of their way to "keep company" with their males, especially when those males are particularly attractive. Among blue tits, for example, attractive males are followed more by their mates than are unattractive males, with "attractive" defined as "those males that receive many visits from neighboring fertile females." Presumably, female blue tits whose mates are unattractive are confident that their males will not be "hit upon" by other females; in addition, such females may well be spending some of their free time prospecting for their own EPCs from other, more attractive males!

There is another reason why females occasionally interfere with their mates' EPC endeavors. To some extent, natural selection is a zero-sum game: Success for you means less success for me and vice versa. This is especially true if certain key resources are in short supply: If, for example, an environment can only support a restricted number of youngsters of a given species, then the offspring of female A will do less well in competing with the offspring of female B if B has been inseminated by a high-quality male. Thus, a female mated to such a male could be more successful if his good genes weren't spread around. This is because her own offspring would not have to compete with others who share their high-quality genetic endowment. By mating repeatedly with an especially desirable male, a female could enhance the competitiveness of her own offspring by keeping the high-quality stud functionally depleted of sperm—and thus unlikely to fertilize other females.

It is notable that female mate-guarding—whether it leads to female–female aggression or enhanced sexual solicitation of the male—has been observed among polygynous species but only rarely among socially monogamous ones. This is counterintuitive. It would seem that possible loss of the male's assistance would be a greater blow to a socially monogamous female (who is more likely to rely on her mate's assistance) than to a polygynous one, who presumably is already resigned to sharing her male with others in his harem. Apparently, it is rare for socially monogamous males to follow up an EPC with other behavior that is costly to their mates; by contrast, among polygynous species, males are more likely to help secondary or tertiary females, or even to invite them into the harem.

In any event, the use of sex as a mate-guarding strategy is not limited to birds (or possibly humans). It may occur in lions, where females in heat are known to solicit—and obtain—upwards of 100 copulations per day during a stretch of four or five days! When it comes to copulations, the lion's share is truly impressive. But one must ask "Why bother?" After all, in the case of lions, there is no question of females enforcing monogamy on the males, since lions typically live in prides consisting of many females, sometimes six or more, and it seems unlikely that her reproductive physiology is so inefficient that a healthy lioness must copulate every 10 or 15 minutes for days on end just to get fertilized.

The answer seems to be that the king of the jungle and his descendants must often endure periods when prey is scarce, and at such times, cubs in particular are at great risk of starvation; accordingly, it is at least possible that by making extraordinary sexual demands on the dominant male, a lovelorn lioness makes it unlikely that another female will be fertilized at the same time. The likely outcome is that when her cubs are born, they will not have to compete with another litter, born to another female. If so, then natural selection would reward lionesses who are the most sexually demanding, resulting in a kind of serial reproduction: not monogamy, but a system in which even polygynously mated females reproduce one at a time rather than all at once.

The suggestion has been made that multiple mating by females may be tactic of nonhuman primates as well, designed to deprive other females of sperm from their sexual partner. After all, even though sperm are cheap, they are not infinitely replaceable, and even the "studliest" of males may have difficulty producing a constant and undiminished supply. It is even possible that something akin to female–female competition for male sexual attention explains an interesting womanly mystery: menstrual synchrony. It is a well-known fact that when women live together—in dormitories, sororities, rooming houses—their menstrual cycles tend to become synchronized. Young women typically begin the academic year with their periods randomly distributed throughout the calendar, but by finals in May or June, nearly everyone in the same domicile is reaching for tampons on the same days.

It is, as Yul Brynner famously put it in *The King and I,* "a puzzlement." But not an impossible puzzlement. Maybe by "agreeing" to ovulate at the same time, women are revealing an ancient, prehistoric, and adaptive response to primate polygyny, whereby they reduce the ability of one male to monopolize the fertility of many different women. One problem here is that it is difficult to see how this strategy would benefit subordinate females; they would seem to be better off if they refused to play along and didn't synchronize with the cycles of dominant females. But perhaps they have little choice—perhaps their physiology is simply manipulated by the latter. Moreover, even this could ultimately be in their interest, if the offspring of less-dominant females would be subjected to damaging competition when resources were limited and if their birth coincided with that of infants conceived by dominant females.

If such strategizing seems far-fetched, bear in mind once more that it needn't be conscious at all and is no more sophisticated than the astounding "strategies" by which liver cells deactivate toxins or nerve cells conduct impulses. And, in fact, it isn't only among the "higher" vertebrates that females are known to compete by sexually monopolizing a male. Consider the so-called poison-arrow frogs, named because lethal chemicals contained in the skin of brightly colored individuals are used by indigenous rain-forest peoples to envenom their arrows and darts. Among many of these amphibians, males provide most of the parental care, so it is in a female's interest to restrict his sexual attention to herself. In such cases, females often remain near their mates, engaging in repetitive courtship behavior, probably to prevent them from mating with other females.

Caring for offspring emerges time after time as a key issue in the maintenance of monogamy. Most biologists' ideas about the evolution of monogamy have long centered around the presumed necessity of male parental care. The idea was that females nearly always prefer monogamy because it gives them—and their offspring—the undivided attention of a caring male. According to this reasoning, polygyny becomes possible when males can be "emancipated" from parenting duties; that is, when females are capable of carrying the whole burden by themselves. (Also, of course, when females aren't too aggressive toward each other ... most likely when they don't need their male's help in rearing offspring.)

There is probably more than a grain of truth in this assumption. Thus, even though monogamy is largely a myth, even among birds, it remains true that the avian world is more prone to monogamy than is any other group of animals. Not coincidentally, birds have very rapid metabolisms and nestlings typically must be fed immense quantities of food, sometimes an insect every 15 seconds or so! Given such extraordinary demands, there is an obvious payoff in having two committed adults caring for a brood, so it is understandable that social monogamy is something of an avian specialty.

For the same reason, it is understandable that monogamy is especially rare among mammals, since female mammals are uniquely qualified to nourish their offspring. Male mammals—although not altogether irrelevant—have comparatively little to contribute. When males do provide for their offspring, they must—not surprisingly—be assured of fatherhood; that is, males are likely to behave paternally only when females do not engage in many EPCs. Hence, we find remarkable paternal solicitude in ostensibly monogamous species such as foxes, beavers, and certain nonhuman primates such as the tiny New World marmosets. In some marmoset species, males carry the juveniles almost as much as females do, and they even act as "midwives," assisting at the birth of their offspring, who—not coincidentally—are likely to be really theirs.

The predominant theoretical explanation for polygyny, the "polygyny threshold model," proposes that females elect polygyny when, by doing so, they obtain sufficient additional benefits (food, good nest-sites, protection) to make up for the undiluted parental assistance they would otherwise receive from a monogamous mate. On the other hand, this model was developed with birds in mind, and it remains ornithocentric. When it comes to direct paternal care, male mammals generally have less to contribute than their avian counterparts; a would-be polygynist can frequently offer a female more in resources, territory quality, and so forth, than she would lose if she forgoes his paltry paternal assistance. And so most mammals are not even socially monogamous.

Human beings are a dramatic exception to this generalization. Baby people are more like baby birds than like baby mammals. To be sure, newborn cats and dogs are helpless, but this helplessness doesn't last for long. By contrast, infant *Homo sapiens* remain helpless for months . . . and then they become helpless toddlers! Who in turn graduate to being virtually helpless youngsters. (And then? Clueless adolescents.) So there may be some payoff to women in being mated to a monogamous man after all.

It is an interesting proposition that monogamy may have arisen as a male response to sperm competition; that is, as a way for males to minimize the risk that someone else's sperm will fertilize the eggs of a given female. Male rats, for example, prefer the odor of unmated females to that of females that have recently mated with another male. It seems likely that polygynous males are more liable to being cuckolded than are monogamous males, simply because it is harder for a male to keep track of many wives than of just one. This, in turn, could help tip the balance in favor of monogamy or, at least, ostensible monogamy. The argument may seem contradictory, but it is internally consistent: Females, as we have seen, have many reasons to seek EPCs, and, moreover, they have become quite adroit at hiding such behavior from males. Male animals, unlike the fabled sultans of yore, cannot enlist the assistance of eunuchs to guard their harems. Accordingly, it is quite possible that such males find that in the long run their reproductive success is higher if they have only one mate, and keep close tabs on her extracurricular sex life, than if they accumulate many . . . each of whom might be unfaithful to him.

This speculation can be carried further, to the possibility that mate-guarding (by either sex, or both) in turn paved the way for the evolution of paternal care. The idea is that if a male remains closely associated with a female, copulating with her until she is no longer fertile, then he has a high level of confidence as to his own paternity and thus is predisposed to help care for the offspring. Moreover, in the case of birds at least, he is likely to be on the scene when the eggs are laid and is therefore available to do his share of the parenting. This is the case among birds because it is most common for one egg to be laid per day and for that egg to be fertilized shortly before it is laid. If a male bird copulates with a female just before she lays her egg, such an egg will probably be carrying his genes. In birds, therefore, there is a payoff to being the last male to copulate, and since one egg is typically produced per day, there is a further payoff to staying around; hence—

perhaps—monogamy. Among mammals, on the other hand, the benefit of male attentiveness is generally more limited, since estrus is more restricted in time. It therefore pays male mammals to be sexually attentive to a female when she is in heat, but—unlike in the case of male birds, who are in attendance at egg-laying because they are likely to be have copulated with the female immediately before—there is no particular reason for male mammals to be present when the female is giving birth.

Unlike birds, mammals experience a long delay between copulation and birth, during which time the female is pregnant. It therefore doesn't seem likely that fatherly behavior among male mammals simply results from the fact that they are nearby when "their" female gives birth, as has been argued for birds. (Remember, daddy birds may be present in the avian delivery room simply because they've recently copulated with their female.) At the same time, it seems that when mammals *are* paternal, it is when they have been in attendance not just at the birth of their young but also throughout their mate's estrous; after all, female mammals, unlike insects and birds, lack obvious sperm storage organs. Paternal mammals are likely to be fathers, not cuckolds . . . a correlation unlikely to be coincidental.

Often, mammalian monogamy and paternal care depend on the alternatives available. Earlier, we considered those large ground squirrels known as hoary marmots—western, mountain-dwelling relatives of the woodchuck. They show a range of male parenting behavior, from devoted fatherhood to virtual indifference. Local ecology holds the key. Some marmots occupy large open meadows in which there are numerous adult males, many adult females, and a range of juveniles. Under these conditions, males spend much of the early summer gallivanting about in search of additional mating opportunities and also mate-guarding, attempting to thwart the gallivanting of other males. Males in this busy social situation essentially ignore their offspring; they are too taken up with the press of sexual threats and opportunities. On the other hand, some marmots occupy small isolated meadows that can only support what is essentially a single extended family: one male, one or two females, and their offspring. Here, without the social and sexual distractions of other adults, male marmots settle down and become devoted family men, playing with their offspring, warning them of the approach of predators, and so forth. It may be significant that under these more isolated conditions, males are also more likely to be the fathers of those young toward whom they are so solicitous.

In the case of hoary marmots, as for most mammals, it appears that males are *capable* of paternal behavior; it's just that they can often reap a more substantial payoff by interacting with other adults . . . especially because their mates are guaranteed to lactate and, thus, to provide at least minimal care. Substantial parental involvement is evidently a lower priority, something that happens only by default: "If there are no other females to solicit, and no other males to worry about," one can almost hear the males announcing to themselves, "then I may as well help take care of the kids."

In contrast to male parenting, which is generally "facultative"—that is, something that may or may not happen—female parenting (especially among mammals) is likely to be obligatory. Even female birds generally get stuck with most of the child care. Among birds known as lapwings, both monogamy and polygyny occur, and yet neither males nor females behave differently from one social situation to the other. Whether monogamous or polygynous, female lapwings still end up bearing the brunt of parental care. When mated polygynously, females end up doing nearly all the incubation, to the extent that they have little time available for feeding themselves. When mated monogamously—so that, presumably, they had the advantage of male assistance with some of the parenting chores— things are no better!

(Something distressingly similar seems to happen in at least one other species; namely, *Homo sapiens*. Even in supposedly liberated households, women are often expected to do most of the domestic and child-related chores. Even when women work full time, men typically do very little to pick up the slack at home. For a large number of women, vocational effort is simply added to home effort.)

Ecological circumstances seem to loom large in the "decision" whether a species—or a particular individual—is monogamous, polygamous, promiscuous, or anything else. Examples are as diverse and as fascinating as life itself. Here is a small sampling: The birds of paradise, a group of tropical species in which the males sport spectacular plumage—hence their name— are nearly all polygynous, with males seeking to mate with a relatively large number of females and providing essentially no parental care. However, in a closely related species, the trumpet manucode, males are faithful mates and devoted fathers. The reason? Manucodes eat figs, which are rich in carbohydrates but comparatively poor in protein and fat. In addition, fig trees are relatively rare (at least in the mountains of New Guinea, where these birds were studied) and, moreover, they fruit unpredictably. Two parents are needed, therefore, to bring enough of the low-quality, semidigested fig glop to the nestlings to assure their survival. The result? Manucode monogamy— not from manucode morality, but fig-forced. Among the other birds of paradise, incidentally, the preferred food is more energy rich, and so females can generally provide enough by themselves, which in turn emancipates the males to go seeking additional sexual partners . . . which they do.

One of the attractive things about monogamy is its egalitarianism: one male, one female, everyone equal. After all, polygamy is inherently unequal, regardless of which sex ends up being the harem-keeper. If one male "has" many females (polygyny), there is the implication that he is somehow "worth" all of them put together and is more valuable and more important than any one of his wives; similarly, if one female "has" many males (polyandry), it seems that the worth and value of each of her husbands is less than that of the one, dominant female. Not so with monogamy, which—whatever its biological difficulties—feels morally right, if only because it is a perfect tie, a 50–50 compromise in which both male and female enjoy equal weighting.

It is also appealing to think that monogamy is a situation of mutual gratification in which the needs of male and female are equally met. This can in fact be true, but a darker, more cynical view can also be justified, one in which monogamy is not so much the result of a beneficent positive-sum game, with balanced winners all around, as of a prolonged conflict of interest in which both males and females lose equally. (Instead of a glass half full, monogamy may thus be a glass half empty.)

The fundamental conflict here is the interest of each sex in confining its mate to only one sexual partner—itself—while at the same time obtaining additional mates, via either polygamy or EPCs. Mate-guarding would therefore be one of many examples of how such conflicts of interest play themselves out. It is most conspicuous on the part of males, each of whom typically seeks to enhance his fitness by reducing his mate's opportunities to engage in EPCs . . . contrary to the female's preference. So, as we have seen, females occasionally circumvent mate-guarding, sneaking away for EPCs when possible. When females mate-guard, their goals are similar: to prevent their male from obtaining those additional mates—or matings— that he would prefer. In this admittedly glum perspective, monogamy is still a success story of sorts: Each sex is equally successful in thwarting the other's desires!

Close and persistent mate-guarding can be costly in other ways—as demonstrated in animals as distinct as dunnocks (birds) and water striders (insects)—interfering with the females' ability to obtain food or avoid predators. Sometimes, as in the water strider case, there are direct energy costs, since guarding males commonly mount the females, who are then obliged to carry their "guardians" on their backs.

The opportunities for conflict seem endless. For example, there may be conflict over resources provided by the male. Among scorpionflies, males generate a salivary mass rich in calories that is transferred to the female during mating; while the female munches, the male mates. The larger the male's nuptial gift, the longer the duration of copulation (simply because it takes the female longer to consume it); and the longer the duration of copulation, the greater a male's success in fertilizing the female's eggs. Also, the larger the male's salivary mass consumed by the female, the greater the size and success of the eggs that the female is able to produce. Not surprisingly, when a male's salivary mass is too small, females seek to end the copulation . . .

prematurely, from the male's perspective. At this point, the male scorpionfly is likely to use a specialized anatomical structure, his notal organ, which clasps the female—evidently against her will—and coerces her to continue mating longer than would be in her interest.

The natural world is filled with similar examples of the male–female conflicts that are inherent in sexual reproduction. Among fruit flies, for example, males benefit from repeated matings. The more copulations, especially with additional females, the more offspring. But for females, the situation is quite different. Although a minimum number of matings is necessary if they are to reproduce, matings are costly. In fact, copulation is dangerous to their health: The more a female fruit fly mates, the shorter her life span. This is because, during insemination, males introduce not only sperm but also a sublethal cocktail of chemicals that enhance their own reproductive success, but at a cost to their female partners. These sexual substances induce females to lay eggs more rapidly, diminish their receptivity to other males, and do battle with any sperm already present in the female's reproductive tract. These latter chemicals in particular appear to damage the female and shorten her life span. All this male–male chemical warfare may also reflect yet another devious male strategy: by producing such a nasty brew, they not only compete with prior (and subsequent) males, they also might restrict the eagerness of females to mate yet again with another male, since any who do so will obtain yet another dose of troublesome chemicals. The result is not quite male-imposed female monogamy, but doubtless a female disinclination to be as sexually adventurous as they would otherwise be.

One notable study, conducted by William Rice of the University of California at Davis, sought to tease apart the likely adaptations and counteradaptations of male and female *Drosophila* by establishing a system in which one sex was, in effect, unilaterally disarmed. Through complex genetic manipulations, male *Drosophila* in a laboratory population were permitted to evolve, but females were prevented from developing counteradaptations. After 41 generations, the evolving male line was more successful in sperm competition than males from control populations, whereas the nonevolving female line suffered unusually high mortality due to the toxicity of seminal fluid. Under normal conditions, when natural selection acts on females as well as males, the tug-of-war between male adaptations for sperm competition and female counteradaptations evidently results in a more balanced outcome, with each kept in check by the other.

Sexual conflict generally occurs when the reproductive success of one sex is enhanced by something that reduces the reproductive success of the other. Males accordingly seek to minimize sperm competition, as by mateguarding, whereas females try to promote it. Among dunnocks, an alpha male mate-guards to protect his paternity, whereas a female attempts to escape the guarding alpha male and actively encourages copulation attempts by beta males, probably so as to induce the beta male to assist with rearing her chicks (by increasing the probability that they might also be his).

Among mammals, we have considered the possibility that reproductive synchrony may be a female counteradaptation to polygyny, making it more difficult for a single harem-keeper to inseminate—or even to guard—all the fertilizable females. Insofar as synchrony enhances opportunities for females to engage in EPCs, it is likely to be in the interests of females but counter to the interests of males.

Then there is the matter of copulatory plugs: Males and females have a shared interest in preventing a male's sperm from leaking out. (This assumes, of course, that the inseminating male was chosen by the female in question and did not coerce the copulation.) But perspectives probably differ on whether it is desirable to keep other males from getting in! Hence, it may be worth noting that female fox squirrels as well as eastern gray squirrels have been observed removing copulatory plugs from their vaginas—either discarding or eating them—within 30 seconds of copulating. Male squirrels would presumably prefer their sexual partners to leave these plugs in place, but there is probably little they can do about it.

In many animals, males prefer to mate with virgins. This way, they gain a higher confidence of paternity. In addition, among species such as the scorpionflies in which males provide a metabolically valuable nuptial gift, a male whose partner is engaging in sex for the first time is guaranteed that his investment will go toward nourishing eggs fertilized by him rather than by some preceding male. But females are likely to see things differently, since they can gain extra nutrients by mating with more than one male. Among bush crickets—another insect species—males cannot distinguish virgins from nonvirgins, because females have evidently evolved means of disguising chemical and physical cues as to their erotic history. Not to be entirely outdone, however, male bush crickets have adopted another tactic to assess the sexual desirability of would-be mates: They evaluate females by age, preferring those that are younger, since all other things being equal, younger individuals are less likely to have already mated. (Of course, a preference for virgins—as marriage partners, as opposed to one-night stands—is not limited to nonhuman animals. Among many human societies, it is still considered a major transgression for a woman to have lost her virginity before marriage; indeed, there is a cross-cultural tradition that voids a marriage if the bride is not a virgin. Physicians in Japan and—more covertly—in the Middle East have long had a booming business re-creating virgins via plastic surgery.)

When it comes to male–female reproductive conflict, there is a pattern that goes even beyond the infamous double standard; namely, sexual coercion. Coercive sexual practices are generally one-way affairs, imposed by males on females, with rape only the most extreme form. And rape has few if any implications for monogamy. By contrast, there is room for members of one sex—usually, but not always, males—to induce females to cease their wandering ways and accept monogamy in return for certain promises, generally promises involving parental care.

Biologist Patricia Gowaty introduces the novel and milder concept of "helpfully coercive" males, individuals who manipulate females into mating with them by helping the females rear offspring. In some cases, the benefits of such help may not be sufficiently great to make up for the lost genetic benefits potentially obtained via EPCs, and females will resist sexual monogamy or else follow this time-honored strategy: Enter into social monogamy but engage in EPCs on the sly.

Those females with a lot to lose if the male withdraws his assistance are most likely to be coerced (or, to put it more kindly, seduced) by male offers of parental helpfulness. Similarly for those females with relatively little to gain via EPCs; that is, if there are only minor differences in genetic quality among available males. It is also possible that high-quality females are especially able to succeed in child-rearing without male help. If so, then these high-quality females—being less fearful of losing male assistance—should be especially liberated to seek extra-pair sex. This, in turn, might provide a novel explanation for why broods with EPC-produced young often have a higher success rate: not so much because extra-pair males provide especially good food, genes, protection, or anything else, but simply because these broods are produced by "superfemales," who, buttressed by their superiority, are capable of disregarding the helpfully coercive efforts of their husbands to keep them at home.

Not that males don't typically try to keep their mates in thrall, often indirectly. Males who compete successfully for choice real estate, for example, may be doing so in an effort to prevent their eventual mates from looking elsewhere for mating opportunities. As a result, it is possible that such females are thereby seduced into mating with males of relatively low genetic quality. But this is unlikely, since, in all probability, high-quality males—because of their high quality—are successful in male–male competition for resources, too. Pair-bonds can, nonetheless, become pair bondage, even without direct physical coercion.

If there are no winners in a world of male–female sexual conflicts, it may be reassuring to note that there probably are no clear losers, either. Adaptations beget counteradaptations, measures generate countermeasures, and in any sexually reproducing species, every reproduction represents a precise, mathematically equal triumph for one male and one female. So, although individuals win or lose, males and females overall do equally well. (Technically, males in general are exactly as "fit" as are females in general.) Biologist Leigh van Valen introduced the "Red Queen hypothesis" as a new evolutionary law: Recall the scene when the Red Queen of Wonderland instructs Alice that with the world moving so rapidly, everyone must run just to stay in place . . . and to get anywhere, it is necessary to run twice as fast! The Red Queen hypothesis states that for systems—such as males and females—that are inextricably tied together, "winning" by one side necessarily generates a compensating response by the other. No matter how fast or how far we run, we are all in the same boat.

Next: human beings. As we shall see in the next chapter, monogamy is not natural to our own species. And indeed, it is much less common than a naive, sentimental view of "marriage and the family" would suggest. But it does occur, and we must ask why.

The short answer is that no one knows. There has, however, been a lot of interesting speculation, not all of it from biologists. In *The Origin of the Family, Private Property and the State*, Friedrich Engels (best known for his collaboration with Karl Marx in writing *The Communist Manifesto*) developed one theory, traceable at least to Rousseau and perhaps to earlier origins. In the beginning, according to this speculation, there were no exclusive social or sexual relationships. Children belonged to everyone. Then came the initial worm in that Edenic apple: private property. Whoever had it wanted to pass it on to his descendants, but with "omnigamy"—everyone copulating with everyone—whose children were whose? The solution was monogamy, whereby men could control the sexuality of women, thereby insuring heirs and validating their property interests.

Engels missed some important points. First, *why* would men be concerned with establishing property rights for their heirs at all? (Because of the deep-seated biological inclination to favor one's genetic relatives, or, to put it in "selfish-gene" language, because genes tend to favor copies of themselves in those bodies that we call "children"?) And second, both Engels and Rousseau before him almost certainly got it backwards: Men don't dominate women so as to protect their property. More likely, they accumulate property—as well as prestige and power of other sorts—so as to attract women.

There have been several other theories to account for the origin of human monogamy. Another nonbiological one was developed by pioneer sociologist Thorstein Veblen, in *The Theory of the Leisure Class*. Veblen attributed much social behavior to a need for self-display, including the accumulation of women. For a man to "have" a woman—and better yet, many women—is a clear sign that he is powerful and successful. And in turn, according to Veblen, "the practice of seizing women from the enemy as trophies gave rise to a form of ownership-marriage." Women were primordial property.

Other ideas reflect more evolutionary sophistication. Thus, we have already considered how the rapid growth of young birds—typically combined with the utter helplessness of many nestlings—seems correlated with avian monogamy . . . such as it is. Human infants grow more slowly than infant birds, but they are, if anything, even more helpless at birth, and they remain profoundly needy for a very long time. Under these conditions, it is reasonable that mothers, and even fathers, would be predisposed to shared parenting, quite literally for the sake of the kids.

The likely connection to monogamy would doubtless go beyond a shared male–female interest in cooperative parenting: To some extent, males can be expected to insist that in return for fathering—and not mere inseminating—they are offered a high confidence of paternity. And, in fact, although human beings are perfectly good mammals, we are also unusual in our fathering.

Aside from pregnancy and lactation, there is nothing that mothers do that fathers can't. Nonetheless, even though paternal care is better developed and more frequent in human beings than in most other mammals, it is not all that prominent even in our own species. A survey of paternal behavior in 80 different societies reported a close father–infant relationship in only 4 percent of them. And even in these cases, fathers spent less than 15 percent of their time actively engaged with infants. Moreover, when they do interact, much of what transpires is play rather than caretaking as such. On the other hand, even though fathers consistently do less fathering than mothers do mothering, there is evidence that what they do may be quite important. Among the Ache Indians of South America, for example, a child is relatively safe and secure if he or she has a father: Up to age 15, such a child suffers only a 0.6% chance of dying. On the other hand, a child under age 15 and lacking a father suffers a 9.1% risk of death.

Given that there can be a real payoff to having a dedicated male caretaker, it is not a great stretch to consider that there may also be a benefit—to the offspring, and thus to both parents—in monogamy. After all, monogamy means that a male, no less than a female, is available for child care. On the other hand, if the issue is protection and access to resources, then it stands to reason that a powerful, wealthy polygynist could in theory provide as much to each of his many wives and their offspring as could a less impressive, impoverished man, even if the latter is devotedly monogamous.

Monogamy is widely seen as benefiting women, while it is often assumed that polygyny is a patriarchal, male-dominated system that oppresses women. But it is easy to forget that for every successful polygynist man, there are several unsuccessful bachelors; since there are roughly equal numbers of men and women, if one man has ten wives, for example, then there are nine without wives at all. Focusing on the high achievers is reminiscent of those people who claim to remember their previous lives: For some reason, their recollections always involve a prior existence as Napoleon, Joan of Arc, one of the pharaohs or queens of Egypt, or a medieval lord or lady . . . never as a serf, slave, or peasant!

It may be that, in fact, polygyny is a disaster for most men and, comparatively speaking, a good deal for most women. With polygyny, more women have the option of associating themselves with a powerful, successful man. For subordinate, less successful men, it is a serious problem, but very few women are likely to be shut out. Thus, although we often think of monogamy as benefiting women, it may be far more congenial for *men,* especially those in the middle or lower ranks. Monogamy is the great male equalizer, a triumph of domestic democracy.

On the other hand, even systems of ostensible monogamy—as found in modern Western societies—often allow successful men, married or not, to have additional sex partners, a situation typically not condoned for married women. There are also other ways of maintaining the legal form of monogamy while circumventing it nonetheless. Of these, the most common pattern is serial monogamy, in which powerful or wealthy men divorce their wives and take up with younger (more fertile and physically attractive) women as their earlier wives grow older. The movie *The First Wives Club* depicted the situation of women deserted by husbands seeking younger spouses as their financial success enabled them to accumulate a kind of harem, albeit serially instead of simultaneously. The question arises: Is serial monogamy, wherein powerful, successful men abandon their wives, more humane than polygyny, wherein they might simply add additional mates? (Serial monogamy is also practiced, on occasion, by wealthy and successful women, and although it appears to be less frequent than its male counterpart, no data are available.)

To be sure, various coercive causes of human monogamy are also less than humane. There is, for example, the fear of social ostracism or—in the case of traditional Roman Catholics—fear of excommunication if they elect to terminate a marital union considered sacred by the Church in order to marry someone else. There can also be fear of physical injury inflicted by a "wronged" spouse: The most common cause of one spouse killing another is sexual jealousy, specifically a man's suspicion that his mate has been unfaithful. One must wonder how many marriages are kept together by fear.

Aside from fear of death, battering, ostracism, and damnation, there is also fear of abandonment, poverty, and vulnerability. One must also ask, accordingly, how many marriages are kept together by a woman's fear that if she leaves her husband—especially if she has dependent children and may also have sacrificed her own economic prospects for her husband's—she is liable to have a terribly difficult time making ends meet, particularly if she is no longer young and attractive. The alternative is a kind of long-standing sexual exchange, almost an inverted prostitution, whereby women exchange fidelity—or, at least, the appearance of fidelity—for resources, especially a decent standard of living and protection. When it comes to the evolution of monogamy, however, the phenomenon to be explained is less female fidelity than male willingness to "stop at one." (After all, part of the burden of this book is that fidelity—whether male or female—is more myth than reality.)

And, finally, there is the old chestnut: staying together for the sake of the children. It sounds trite, but is nonetheless real for thousands, probably millions, of couples. Insofar as the benefit of shared parenting is a major cause of monogamy, it may be altogether understandable—if profoundly unromantic—that this benefit is often responsible for maintaining monogamy as well . . . for better or worse.

Kristen Hawkes, an anthropologist at the University of Utah, reports that there are many isolated pretechnological societies in which paternity is unclear, suggesting that maybe men get something else from monogamy. Among the Bari of Colombia and Venezuela, "partible paternity"—the idea that a child can have several fathers—is a common belief: 24 percent of Bari children and 63 percent of Aché children (Paraguay) were said to have multiple fathers. And kids were better off with multiple fathers than with just one: 80 percent of the former survived to age 15 versus only 64 percent of those with a "single father." This finding also supports Sara Hrdy's theory, described earlier, that female sexual receptivity to multiple males (and concealed ovulation) is a strategy to keep males uncertain as to paternity. Hrdy sees this as a way of avoiding infanticide, but it might also help generate tendencies to provide dependent children with food and other crucial resources, as well as defending them if need be.

Hawkes also points out that among the Aché of Paraguay and the Hadza of northern Tanzania, different families are likely to receive equal portions of meat brought back from the hunt. This, also, doesn't fit the "bargain hypothesis" of monogamy, under which a successful hunter's wife—and her offspring—should get to hog the proceeds. Hawkes has found, however, that successful Hadza hunters have younger wives, have more EPC partners, and father more children than do the less successful hunters. Their offspring also have a higher survival rate, perhaps because of better nutrition or because successful hunters got to select more competent women as wives (and, thus, mothers). According to Hawkes, monogamy may have arisen as a result of "negotiations among males," whereby access to women is divided up and harmful fighting is avoided.

Nonetheless, as we shall see in the next chapter, there is overwhelming evidence that monogamy is not "natural" to human beings. If, then, monogamy is a rather peculiar and derived state, why has it been victorious . . . at least in official doctrine, if not in the actual observance? A simple answer is that it is the avowed marital system of Western countries, whose military, economic, and cultural power have simply imposed Western preferences on the rest of the world. But this ducks the question: Why is monogamy approved—in theory if not in practice—in these Western countries?

To some extent, it may be a little-appreciated example of the triumph of democracy and "equal opportunity," at least for men. Polygyny, as already mentioned, is a condition of elitism, in which a relatively small number of fortunate, ruthless, or uniquely qualified men get to monopolize more than their share of the available mates. With monogamy, by contrast, even the most successful individual cannot have more than one legal mate; as a result, even the least successful is likely to obtain a spouse as well. If it requires some suppression of our tendencies for multiple mates, monogamy offers in its place an enhanced prospect of there being a mate for each of us.

The early evolutionary history of human mating systems is unclear, but here is a possible quick-and-dirty scenario: Our early ancestors roamed the Pleistocene African savannas in small bands. Most current hunter-gatherer human societies are monogamous (although with adultery relatively common). Only a small number of men—typically 5 to 15 percent—are active polygynists, and even then, it is extremely rare for one man to have more than a few wives. With a hunter-gatherer lifestyle, it is almost impossible for one man to obtain a monopoly of resources, or even a preponderance: Luck is often involved in hunting, for example. Besides, when game and gathered vegetable foods are at issue, it is difficult to store excess. Then came agriculture, providing the opportunity for some men to own large amounts of land and, with it, large surpluses. The rich were able to get richer yet. Out of this came enhanced competition as well as the prospect of enhanced success—especially more wives—for the winners. In a sense, maybe Rousseau wasn't altogether wrong, after all, when he suggested that people were primitively egalitarian, with this Eden disrupted by the invention of private property!

In any event, with increased wealth concentrated in the hands of a few came the prospect of additional wives. Polygyny flowered—and not just among early farmers: The same applied, if anything more so, to pastoralist societies, nearly all of which were traditionally polygynous. Even today, large herds of cattle, goats, camels, and so forth equal large wealth, which equals large numbers of wives for the "haves."

Polygyny was evidently widespread in the Near East, as reflected by numerous Old Testament references to the many wives of the early Israelite kings. And polygyny continued to be the favored practice in much of the world, especially in what anthropologists and sociologists call "highly stratified societies," those in which there are substantial economic and social differences between the poorest and the wealthiest. As expected, polygyny has long been a preferred system for wealthy, powerful men, continuing up until modern times in India, China, and Africa. Western Europe, however, was a notable exception; even though there has long been plenty of disparity in wealth between the richest and poorest Europeans, these people moved rapidly toward what has been called "socially imposed monogamy."

Opinion is divided as to why. One possibility is that by the Industrial Revolution, most people (i.e., workers) were once again more or less equal in possessions. Of course, successful captains of industry became fabulously wealthy; the point is that even the average wage-slave—although badly off compared to the factory owners—was capable of supporting a family, at least minimally.

Monogamy may thus be, at least in part, a *result* of male–male equality; even more so, however, it is a *cause* of equality, a great reproductive leveler (for men)—at least in biological terms. The possibility therefore arises that, historically, monogamy arose in Europe as an implicit trade-off. The wealthy and powerful would in effect have agreed to give up their near-monopoly on women in return for obtaining greater social involvement on the part of middle- and lower-class men, who, if reproductively excluded, might have refused to participate in the social contract necessary for the establishment of large, stable social units. According to this view, these men were offered a level—or, at least, a somewhat more level—reproductive playing field in exchange for their cooperation in dealing with internal or external threats.

Male–female reproductive conflict, whatever its degree, pales in comparison with the intensity of male–male conflict, which, in turn, may have induced powerful men to accept restrictions (however unwillingly) in order to enlist the aid of their less fortunate but more numerous competitors. And so, instead of bread and circuses, the hoi polloi may have been offered monogamous marriage. In any event, among Europeans in particular, Christianity became especially active in promoting monogamy and imposing it on the secular elite.

The foregoing explanation fails to account for the fact that many highly stratified polygynous societies—such as those of the Incas, Aztecs, Indians, and Chinese—also maintained an extraordinary amount of social cohesion despite an intense degree of non-monogamous reproductive despotism. This criticism is not devastating, however, since the theory relating European monogamy to enhanced social integration simply claims that the connection may be significant, not that it is necessary or exclusive.

Even Bill Gates is legally forced to be monogamous . . . although successful sports and rock stars often have multiple sexual liaisons, and, for all we know, so does Mr. Gates himself. Bill Clinton, too, is legally forced to be monogamous . . . although powerful men are typically inclined to seek additional pairings (if only briefly) and—because of the nature of female sexual psychology—are generally able to find willing partners. It is easy, as well, to imagine queen bees such as Elizabeth Taylor, Madonna, or Oprah Winfrey being in demand—and command—as polyandrous females . . . except for the legal restraints. The point is that although social ideology and legal restrictions cannot change human nature, they can and do impose egalitarianism in several forms: Everyone is supposed to be equal before the law, equally deserving of life, liberty and the pursuit of happiness, and entitled to—or bound by—monogamy.

# CHAPTER SIX

# What Are Human Beings, "Naturally"?

E thel Merman used to belt out a song that advised "doing what comes naturally." (As you might expect, it was about sex.) It is easier to talk—or sing—about what is "natural" than to pin it down, especially when the subject is human beings. After all, people could never, in a sense, do anything truly *un*natural: We cannot survive without air, walk upside down on the ceiling, or grow additional heads. At the same time, everything we do, everything we are, is a consequence not only of our internal "human nature" but also of our experiences. There is no way to know for certain what human beings would be like without the influence of the environment, since a person without an environment could not survive, not to mention behave in interesting and meaningful ways.

Nonetheless, people keep trying to tease apart the roles of experience, environment, culture, social tradition, and so forth, hoping to get a glimpse of what "naked, unaccommodated man" (as Shakespeare's Lear put it) is like. The devout King James of England—who commissioned what became the most famous translation of the Bible—is said to have arranged for a few children to be reared in complete isolation from any spoken language, so as to ascertain the "natural" human language, uncontaminated by the prompting of others; James apparently hoped that the unfortunate subjects would spontaneously begin speaking Hebrew! They didn't. (Not even Aramaic.)

Nonetheless, the search for bona fide human "naturalness" goes on. Underlying much of the research of anthropologists has been the unspoken hope that by studying the lives of nontechnological people, they would gain insight into the "natural," unaccommodated inclinations of *Homo sapiens,* unpolluted by MTV, computer games, air conditioning, or microwaved popcorn. Even though such efforts can easily be caricatured, there is nonetheless something meaningful about asking what human beings, deep inside, really and truly, unconstrained by civilization—or at least, minimally constrained—naturally *are.* To be sure, no one lives in a pure state of nature, but if we look both intensively and extensively at human beings, it is possible to discern some patterns that persist.

One such pattern concerns monogamy, polygamy, and EPCs.

Twenty-five hundred years ago, Plato offered an answer—albeit tongue-in-cheek—to this chapter's title question. His dialogue *The Symposium* presented this scenario, put in the mouth of the bawdy playwright Aristophenes: It seems that originally people were not as they are today, divided into male and female. Rather, they were happy, fulfilled, self-contained Androgynes, each individual consisting of four arms, four legs, two heads and sporting duplex genitalia. Each "person" was a self-contained monogamous unit. But Zeus found their smugness intolerable and—as was his wont under such circumstances—commenced flinging thunderbolts, which split the Androgynes asunder. Since then, we have all been doomed to wander the earth, seeking to find our "other half," to reestablish that perfect, prethunderbolt state of united bliss.

(Plato's story, incidentally, has the added charm of providing an explanation for homosexuality as well as for heterosexual monogamy: Not all the original Androgynes, it seems, were truly androgynous. Some were composed of doubled female halves and some, of doubled males. Male and female homosexuals were thus generated by the same process as heterosexuals, and—like heterosexuals—they also seek ultimate satisfaction by reestablishing their preschismatic union.)

According to the Androgyne myth, monogamy—whether heterosexual or homosexual—is our natural state, our route to anatomic as well as emotional wholeness. Seen this way, Plato's tale extols monogamy. Or, alternatively, maybe it does just the opposite: explains the frequent *departures* from monogamy. Thus, maybe monogamy is natural, but only with the correct mate! If one's designated, monogamous union happens to be with the wrong half (who presumably is someone else's correct half), then EPCs are understandable efforts to locate that long-lost partner! ("Are you my missing Other?" "No? Well, how about you?")

But this myth, engaging as it may be, is not the one with which we are presently concerned. Rather, the myth of monogamy that we are seeking to investigate is the opposite of Plato's. It is the more widespread myth, the one claiming that in monogamous union the partners are naturally joined together, like the Androgynes before Zeus's untimely intervention. This myth, the one that not only extols monogamy but also insists on its virtual universality, is a "real" myth; that is, one that is widely held, but false.

As we shall see, the evidence is overwhelming that monogamy is no more natural to human beings than it is to other living things. First, we turn to polygyny.

The evidence includes sexual "dimorphism" combined with sexual "bimaturism." Dimorphism ("two bodies") refers to the fact that males and females are significantly different, not only in their genitals but also in basic bodily attributes, especially size. Even though some men are smaller than some women, and some women are larger than all but the largest men, men are generally larger than women, on average about 10 to 15 percent taller and heavier. (This isn't a value judgment, just a statement of statistical and biological fact.) The most likely explanation for such a disparity is that the larger sex has evolved to be larger because of the payoffs associated with success in competing with other same-sex members. And the most direct route to such a payoff is reproductive success; namely, a harem consisting of more than one female.

Looking at other living things, we find that the greater the degree of polygyny, the greater the degree of dimorphism: Certain rain-forest deer that are essentially monogamous are also monomorphic, while elk—classic harem-keepers—are highly dimorphic, with the bulls considerably larger than the cows. Similarly with members of the seal family: Monogamy and monomorphism go together, as do polygyny and dimorphism. Primates, too: Compare the dimorphic (and polygynous) gorillas, for example, with the monomorphic (and largely monogamous) marmosets.

In some primates, monogamy—or, rather, monandry, female fidelity to one male—seems remarkably persistent, despite opportunities for EPCs. In one study of free-living hamadryas baboons, for example, four out of five males in some isolated groups were vasectomized. After four years, both females associated with the intact male had given birth, whereas of the remaining six females associated with the four vasectomized males, not one had done so. Not surprisingly, male hamadryas baboons have much smaller testes relative to their body size than do the more sexually promiscuous baboons. Human beings are definitely at the big-ball end of the primate spectrum, more like chimpanzees than like gorillas or hamadryas baboons, further suggesting that we have long been accustomed to competing via our sperm as well as our bodies.

Don't be deceived, however, into thinking that human beings can easily be pigeonholed as to their "natural" way of living. Even other species (presumably "simpler" than *Homo sapiens*) can be so variable as to defy ready generalization. For example, in the primitive, egg-laying Australian mammals known as short-beaked echidnas, males form "mating trains" of up to 11 individuals, all neatly—and, to the human observer, comically—lined up nose-to-tail, patiently marching along behind an estrous female, each waiting for his chance to mate with her. But this is only true in the warmer climates of northern Australia. In colder climates (e.g., in Tasmania) echidnas are monogamous. Go figure.

Nonetheless, certain basic patterns can be identified, and, in fact, doing so is much of what science is all about. A Martian biologist, sent to earth to describe its various life-forms, would have little doubt, based on sexual dimorphism alone, that *Homo sapiens* is mildly polygynous.

Added to this is the evidence from *sexual bimaturism,* the peculiar phenomenon whereby girls become women a year or two before boys become men. Like sexual dimorphism, sexual bimaturism is a consistent, species-wide trait. And like sexual dimorphism, sexual bimaturism has "polygyny" written all over it. Just as being larger and stronger conveys an advantage when it comes to same-sex competition for getting and keeping a harem, there is a payoff in being older, too. So a predictable hallmark of polygynous species is that the males delay their maturation until they are somewhat older, stronger, tougher, and presumably a bit more savvy than their more callow counterparts.

There is more evidence yet.

Men are consistently more violent than women, which again is a predictable trait of the more competitive, harem-keeping sex. A high level of aggressiveness among women carries with it relatively little payoff along with substantial possible cost, since women are unlikely to be rewarded by acquiring a "harem" of men. Biologically, there is relatively little difference between the most and the least successful women; by contrast, there is a huge difference between the most and the least successful men, especially if the mating system is polygynous. This, in turn, results in the harem-keeping sex being not only larger and later-maturing but also, on average, more aggressive. In strictly monogamous species, there is generally very little difference between the sexes when it comes to propensities for violence . . . once more, because there is little to be gained and much to be lost. The harem-keeping sex, by contrast, plays a more high-stakes game and is more likely to use risky strategies; playing it safe makes sense in a world of monogamy or if one is a harem-*member.*

It is tempting to conclude that male–female differences in sexual dimorphism and bimaturism—and even male–female differences in violence and risk-taking—are "hard wired" into human biology. On the other hand, cultural traditions must be acknowledged, especially when it comes to behavioral differences: Society generates expectations that men are "supposed" to act one way and women, another. In fact, even such apparently biological differences as body size and age at sexual maturation could have cultural components—if society dictates, for example, that boys should eat more than girls and thus, perhaps, grow larger. (It is more difficult—although not impossible—to see how age at sexual maturation could be a cultural construct.)

But the idea that such differences are culturally determined founders on this hard fact: They occur *cross*-culturally, in societies as diverse as urban America, the highlands of New Guinea, the arctic tundra, and the islands of the South Pacific. If male–female distinctions were arbitrary productions of culture, then they should vary randomly from one society to the next. But they don't. These differences occur wherever *Homo sapiens* is found, and all of them point to a dollop of primitive polygyny as part of our biological heritage.

There is yet more evidence for an underlying pattern of human polygyny. It comes from a series of studies and observations concerning our species-wide sexual inclinations. Conducted by different researchers at different times, they converge on a few basic principles, including the fact that men, worldwide, have a much greater interest in sexual variety than do women. Once again, this makes particular sense if *Homo sapiens* is inclined to be polygynous, because a harem-keeper will have sexual relations with many women and also because, as sperm-makers, men can increase their reproductive success by doing just this. As a general rule, the male strategy has been to increase the number of sex partners rather than to have more children per partner—that is, to opt for quantity over quality.

In one study, unmarried U.S. college students were asked how many sex partners they would ideally like to have during various time intervals, from the next month to the rest of their lives. Men consistently indicated a desire for more partners than did women: During the coming year, for example, men wanted 6; women, 1. For the next three years, men wanted 10; women, 2. And as time went on, the desired difference in number of sex partners increased, until, on average, men indicated wanting 18 different sex partners over their lifetime, as compared to women's desire for 4 to 5.

Men and women were also asked to estimate the probability that they would agree to sexual intercourse with an attractive member of the opposite sex if they had known this person for one hour, one evening, a whole day, a week, a month, all the way up to five years. Men and women were equally inclined to have sex with such a person after five years, but for every interval short of that, men indicated a higher probability than did women. (For women, the "break-even" point was between three and six months; for men, it was about one week. Women indicated a virtually zero probability of sex after one hour, with essentially no change over the first day; even after one week, women reported themselves as very unlikely to have sex. Men said that after just one day, their likelihood of consenting would be nearly 50–50.)

These studies can be criticized for relying on what people say rather than what they actually do. Here, then, is another innovative piece of research that comes closer yet to measuring actual behavior. An attractive man or woman approached strangers of the opposite sex on a college campus and said, "I have been noticing you around campus. I find you very attractive." Then, they posed one of these three questions, selected for each subject at random: (1) "Would you go out with me tonight?" (2) "Would you come over to my apartment tonight?" (3) "Would you go to bed with me tonight?"

Of the *women* who were asked for a date, 50 percent agreed; of those asked to go to the man's apartment, 6 percent agreed; of those asked for sex, none agreed. Of the *men* who were asked for a date, approximately 50 percent said "yes" (the same as the proportion of women who had consented), whereas 69 percent agreed to go to the woman's apartment and no fewer than 75 percent agreed to go to bed with her that night! Interestingly, among the 25 percent who refused, a large number felt it necessary to explain, pointing to a previous engagement with a girlfriend and so forth.

As with other living things, both men and women are liable to suffer some costs associated with EPCs, such as increased risk of sexually transmitted diseases and other possible physical risks (notably being physically injured by the partner or—if one or both participants are already paired— by one's own mate or the mate of the short-term partner). This latter tariff in particular is most likely to be paid by men. On the other hand, loss of reputation is a significant possible cost that falls most heavily on women: Although men can suffer if they develop a reputation of being promiscuous or unprincipled, the aura of being a Casanova or Don Juan also has its benefits, often enhancing social status among men and even, on occasion, making such a man more attractive to women.

Why? In part because of the well-known consumer effect, whereby any product—shoes, car, food item, Wall Street stock—becomes more desirable if it is seen as being desired by others. And in part because of a logical outgrowth of polygyny as the long-standing, primitive human condition. Polygynous societies are not *uniformly* polygynous; after all, there are approximately equal numbers of men and women, so it is impossible for all men to be harem-keepers! Those who succeed are likely to be especially powerful, physically or mentally well endowed, high in status, and in control of substantial resources (land, domestic animals, other forms of material wealth, social allies). Given this connection, it is not surprising that men who acquire a reputation for sexual access to many women are widely perceived to be high in status or control of resources, so that—by a process similar to that described earlier for the "sexy son hypothesis"—success can lead to success . . . for men.

By contrast, a reputation for promiscuity does far more harm to a woman than to a man. Whereas a man who is sexually successful with many women is likely to be seen as just that—successful—a woman known to have "success" with many men is unlikely, as a result, to have enhanced her reputation. Instead, even in today's more sexually liberated and egalitarian climate, she is more likely to be known as a "slut."

A "fast" woman known to be an "easy lay" may well be popular among men seeking their own short-term sexual liaisons, but not among those looking for a committed relationship. This, of course, is part of the long-lamented double standard, whereby men and women are typically subjected to different sexual expectations. Willard Espy neatly expressed the curious double-bind implicit in the male approach to the double standard:

> I love the girls who don't.
> I love the girls who do.
> But best, the girls who say, "I don't . . .
> But maybe just for you.

Although such standards are undoubtedly heavily influenced by culture, the fact that they are generally cross-cultural—that is, found among a wide variety of human societies—strongly suggests that they are ultimately rooted in biology. Those roots probably sprout from the difference between men and women when it comes to confidence of genetic parentage. (The long term reproductive success of men would not generally be well served by affiliating with women likely to cuckold them; men can therefore be expected to refrain from marrying women who have a reputation for being EPC-prone. By contrast, although a reputation for EPCs could indeed be a detriment to the desirability of a man, especially when monogamy is expected, it would do little to diminish a man's prospects if the expectation were polygyny.)

A woman with a reputation for sexual promiscuity may well be signaling that she is less discriminating, perhaps because she is unable to obtain a high-quality long-term mate. A woman's history of many short-term sexual partnerships may thus have precisely the opposite effect of a similar reputation for a man: It is likely to announce that she is of low quality and minimum long-term desirability.

T he case "for" human polygyny—not as an ethical good but simply as a biological "natural"—is further solidified by the findings of primatologists and anthropologists. Primatologists first.

Monogamy had been reported for 10 to 15 percent of all primate species, as compared to 3 percent or so for mammals generally (and better than 90 percent for birds). As we have seen, reports of avian monogamy—and mammalian monogamy generally—have turned out to be greatly exaggerated . . . like Mark Twain's comment on the rumor of his death. Similarly, field evidence, accumulated after thousands of hours of direct observations of elusive tropical primates in the wild, has been showing that primate monogamy, too, is not what it was cracked up to be. A recent review of the evidence has concluded, in fact, that only nine species of primates live in exclusive two-adult groups throughout their geographic range. (Some others appear to be monogamous under certain conditions only.) So there is no reason to think that human beings represent a mammal group that is unusually predisposed to monogamy.

As to human predispositions, the clearest evidence comes from how people actually lived before the cultural homogenization that came with Western imperialism and the Judeo-Christian ethic of monogamy. (Incidentally, even that ethic was not originally monogamous: King David had at least 6 wives. [2 Samuel 3: 2–5] and later took more [2 Samuel 5:13]. And Solomon is said to have had 700 wives as well as 300 concubines [1 Kings 11: 1–3].)

Students of human society have long been divided as to how and why human beings ever arrived at monogamy. The nineteenth-century Finnish anthropologist Edward Westermarck maintained that monogamy came first and was subsequently embroidered upon as various other marriage systems were developed. By contrast, Lewis Morgan, father of American anthropology, argued that monogamy isn't primitive at all but is, rather, not only an advanced condition but, in fact, the pinnacle of human family structures. In this view, monogamy rests triumphantly, if not chastely, on top, like a man in the missionary position. This view is no longer widely held and has even been ridiculed, as in the following passage from the noted student of South Pacific cultures, anthropologist Bronislaw Malinowski, who noted that, according to Morgan,

> Human society originated in complete sexual promiscuity, passed then through the consanguine family, the punaluan household, group marriage, polyandry, polygamy and what not, arriving only after a laborious process of 15 transformations in the happy haven of monogamous marriage.

Malinowski further notes that, in this view,

> The history of human marriage reads like a sensational and
> somewhat scandalous novel starting from a confused but interesting
> initial tangle, redeeming its unseemly course by a moral denouement,
> and leading as all proper novels should to marriage, in which "they
> lived happily after."

Whatever course the evolution of the human family may actually have
traveled, it is clear that we didn't all arrive at the same place. And, moreover, it is also clear that monogamy was at best a minority destination. Of
185 human societies surveyed by anthropologist C. S. Ford and psychologist Frank Beach, only 29 (fewer than 16 percent) formally restricted their
members to monogamy. Moreover, of these 29, fewer than one-third wholly
disapproved of both premarital and extramarital sex. In 83 percent of the
societies examined (154 out of 185), males were allowed multiple mateships—that is, polygyny or socially approved concubines rather than
monogamy—if they could afford it.

The renowned anthropologist G. P. Murdoch, in his classic study *Social
Structure,* found that of 238 different human societies around the globe,
monogamy was enforced as the only acceptable marriage system in a mere
43. Thus, before contact with the West, on average more than 80 percent of
human societies were preferentially polygynous, meaning that male harem-
keeping was something that most men sought to attain. It is safe to say that
institutionalized monogamy was very rare.

Anthropologist Weston LaBarre concurs:

> When it comes to polygyny, the cases are extraordinarily numerous.
> Indeed, polygyny is permitted (though in every case it may not be
> achieved) among all the Indian tribes of North and South America,
> with the exception of a few like the Pueblo. Polygyny is common too
> in both Arab and Negro groups in Africa and is by no means unusual
> either in Asia or in Oceania. Sometimes, of course, it is culturally-
> limited polygyny: Moslems may have only four wives under Koranic
> law—while the King of Ashanti in West Africa was strictly limited to
> 3,333 wives and had to be content with this number.

The greater the degree of "stratification" in most nontechnological societies, the greater the degree of polygyny. In other words, those who were
very powerful and very wealthy (the two have long been pretty much synonymous) were nearly always (1) men and, also, (2) possessors of large

harems. Not uncommonly, harem size was precisely calibrated to power and wealth. "In Inca Peru, as probably everywhere," notes evolutionary anthropologist Laura Betzig of the University of Michigan,

> the reproductive hierarchy dramatically paralleled the social hierarchy. Petty chiefs were by law allowed up to seven women; governors of a hundred were given eight women; leaders of a thousand got 15 women; chiefs over a million got 30 women. Kings had access to temples filled with women; no lord had less than 700 at his disposal. Typically, the "poor Indian" took whatever was left.

However, a preference for polygyny does not mean that it is always achieved. Even when multiple mateships are considered highly desirable, at any given time most men had no more than one mate. But in these situations, many men, if they lived long enough, still had more than one socially approved mate, since as they grew older, they typically got richer.

So much for polygyny. (Although a great deal more can be said!) For now, the point is that monogamy is under assault from many different directions, one being that it is not the "natural" human condition. On this, the evidence from anatomy, physiology, behavior, and anthropology can be considered decisive. Another point is that even when human monogamy occurs, it is shot through with extra-pair copulations, with the same penchant for EPCs that has been so persuasively documented of late for animals.

In their sample of 185 widely separated human societies, anthropologist Ford and psychologist Beach found that 39 percent not merely tolerated but actually approved extramarital sexual liaisons, generally of specific kinds. Incest was the only consistent sexual prohibition. The Toda people of India, for example, reputedly had no concept of adultery and even considered it immoral for one man to begrudge his wife to another. In many societies, extramarital sex was limited to certain categories, such as brothers and sisters-in-law among the Siriono of eastern Bolivia. These people were "monogamous," but men were permitted to have sexual intercourse with their wife's sisters and with their brothers' wives. Women, in turn, could have sex with their husband's brothers and their sisters' husbands. Among the Haida tribe, married men and women were generally permitted sexual relations with anyone belonging to the spouse's clan; at most, the husband or wife could "object softly." Usually he or she did not. In short, even when monogamy has been the legally instituted form of mateship, it often has not precluded certain specified extramarital relationships, at least among some human societies. Most of the world's peoples, throughout history and around the globe, have arranged things so that marriage and sexual exclusivity are not necessarily the same thing.

In addition to permitting extramarital sex among designated relatives, many otherwise monogamous societies have approved extramarital sex at special times, notably religious or harvest festivals such as the Brazilian Mardi Gras.

Next: polyandry. As a socially sanctioned institution, it is exceedingly rare. It is also a fascinating biological irony that although men stand to gain more—in terms of producing offspring—from multiple copulations, women are physiologically capable of "having" more sex than men. Add to this the peculiar anthropological fact that nearly all social systems are structured the other way around. In his *Letters from the Earth* , Mark Twain had great fun with this paradox. Here is Twain's Devil reporting his discoveries, after visiting our planet:

> Now there you have a sample of man's "reasoning powers," as he calls them. He observes certain facts. For instance, that in all his life he never sees the day that he can satisfy one woman; also, that no woman ever sees the day that she can't overwork, and defeat, and put out of commission any ten masculine plants that can be put to bed to her. He puts those strikingly suggestive and luminous facts together, and from them draws this astonishing conclusion: The Creator intended the woman to be restricted to one man.
>
> Now if you or any other really intelligent person were arranging the fairnesses, and justices between man and woman, you would give the man a one-fiftieth interest in one woman, and the woman a harem. Now wouldn't you? Necessarily, I give you my word, this creature . . . has arranged it exactly the other way.

Twain's Devil is absolutely right: One man is less capable of sexually satisfying many women than one woman is of satisfying many men. Nonetheless, from a biological perspective the difference between eggs and sperm proclaim that it is more logical for one man to mate with multiple women than for one woman to mate with several men. And, in this case, evolutionary logic has won out.

Group marriages are even scarcer than polyandry. Perhaps the most flexible matrimonial system was found among the Kaingang of southeast Brazil: 8 percent of Kaingang marriages were truly group affairs, involving two or more men mated to two or more women; 14 percent involved one woman married to several men; in 18 percent, one man was polygynously mated to several women; and 60 percent were monogamous. But clearly it would be misleading to call the Kaingang "monogamous," even though monogamy was the most frequent pattern. Other arrangements were officially permitted, and, in fact, polygyny was preferred (at least by men).

Even among those societies that can legitimately be described as monogamous—that institutionalize marriage between one man and one woman—sexual relations between husband and wife are less exclusive than a Western Judeo-Christian perspective might anticipate. For example, among avowedly monogamous societies, about 10 percent actually permit relatively free extramarital sexual intercourse. Among the Lepcha of the Himalayas, for example, a husband is expected to object only if his wife has sexual relations with another man in his presence! About 40 percent of ostensibly monogamous human societies permit extramarital sex under special conditions (certain holidays) or with particular individuals (such as the husband's brothers), and only about 50 percent prohibit extramarital coitus altogether. Among these restrictive societies, the rules apply most tightly to wives, much less so to husbands: Only a very small percentage prohibit extramarital sexuality on the part of men.

In a number of human groups, men routinely exchanged sexual relations with each other's wives. Among certain Eskimo, Cumana, Araucay, and Crow Indians, honored guests were permitted to sleep with the host's wife, and the Siberian Chukchee established regularized patterns of wife-exchange, so that a traveler, far from home, could be guaranteed a warm bed and pleasant accommodations. (Nothing has been reported regarding the wives' attitudes toward this system.)

Similar arrangements were set up by the Mende people of Sierra Leone. In this case, wives were supposedly encouraged by their husbands to take lovers; these, in turn, were then expected to provide manual labor to help the family.

Aside from cases in which the rules of marriage allow for sexual relations with persons other than the husband or wife, extramarital sex—even when socially disapproved—sometimes carries only mild penalties; for many of the world's people, it is closer to a misdemeanor than a felony. According to anthropologist Ruth Benedict, for example, the Pueblo people of New Mexico

> do not meet adultery with violence. . . . [T]he husband does not
> regard it as a violation of his rights. If she is unfaithful, it is normally
> a first step in changing husbands, and their institutions make this
> sufficiently easy so that it is a really tolerable procedure. They do not
> contemplate violence.

Wives are often as moderate, if not more so, when their husbands are known to be unfaithful. As long as the situation is not unpleasant enough for relations to be broken off, it is frequently ignored, as with this case among the Zuni:

One of the young husbands of the household . . . had been carrying on an extramarital affair that became bruited about all over the pueblo. The family ignored the matter completely. At last the white trader, a guardian of morals, expostulated with the wife. The couple had been married a dozen years and had three children; the wife belonged to an important family. The trader set forth with great earnestness the need of making a show of authority and putting an end to her husband's outrageous conduct. "So," his wife said, "I didn't wash his clothes. Then he knew that I knew that everybody knew, and he stopped going with that girl." It was effective, but not a word was passed. There were no outbursts, no recriminations, not even an open recognition of the crisis.

The Pueblo are what Benedict termed "Apollonian"—after the Greek god Apollo, deity of the sun, music, medicine, and reason—in that they are reluctant to show violent emotions. Divorce is readily available to the Pueblo people, and, in fact, a wife who remains with her husband after he has had numerous affairs is considered faintly ridiculous, because her per-severance is seen to indicate that she must really love him!

By contrast to the Apollonian Pueblo, anthropologist Benedict described so-called "Dionysian" cultures, in which violent emotion is permitted, even encouraged. For example, on the island of Dobu, off the coast of New Guinea, adultery is frequent, but it is also cause for outrage and jealousy: "Faithfulness is not expected between husband and wife, and no Dobuan will admit that a man and woman are ever together even for the shortest interval except for sexual purposes." The Dobuan husband, according to Benedict, is suspicious even when his wife goes briefly into the bushes to uri-nate. And for good reason:

Adultery within this group is a favorite pastime. It is celebrated constantly in mythology, and its occurrence in every village is known to everyone from early childhood. It is a matter of profoundest concern to the outraged spouse. He (it is as likely to be she) bribes the children for information, his own or any in the village. If it is the husband, he breaks his wife's cooking pots. If it is the wife, she maltreats her husband's dog. He quarrels with her violently. . . . He throws himself out of the village in a fury. As a last resort of impotent rage he attempts suicide by one of several traditional methods, no one of which is surely fatal.

As we shall see, however, even the Dionysian Dobu are mild in their response to adultery compared with many of the world's peoples.

The double standard is widespread in most societies, with men permitted far greater freedom than women to engage in sex outside marriage. After reviewing 116 different human societies, anthropologist Gwen Broude reported that whereas 63 permit extramarital sex by husbands, only 13 permit it for wives. In addition, 13 had a "permissive single standard," allowing extramarital sexual activities equally to both husband and wife, whereas 27 engaged in a "restrictive single standard," prohibiting husband and wife equally from engaging in any extramarital affairs. Similarly, Laura Betzig evaluated the causes of marital dissolution worldwide, concluding that whereas there are many causes—childlessness, economic failures, sexual incompatibility—adultery is "the single most common cause of divorce" and that infidelity by the wife is far more likely to be cited than is infidelity by the husband.

If a female mammal becomes inseminated because of an out-of-pair-bond copulation, she is no less the mother of the offspring produced; but her deceived mate—who may nonetheless provide food, defense, baby-sitting, and so forth—is very much less the father! So it is anticipated that, for most living things, not only will males be more eager for EPCs, but they will also be more intolerant of the same behavior by their mates. The stage is therefore set for the double standard, wherein sexual expectations differ for men and women, as they typically do for males and females of other species.

Friedrich Engels, in *The Origin of the Family, Private Property and the State,* suggested that the human family "is based upon the supremacy of the man, the express purpose being to produce children of undisputed paternity." In a famous oration known as *Against Neaira,* the Greek orator Demosthenes stated the sexist bias of society in his day: "Mistresses we keep for pleasure, concubines for daily attendance upon our persons, and wives to bear us legitimate children and to be our housekeepers." Of these various "uses" of women, the production of legitimate children seems to have been especially important, and it may go a long way in explaining the reason male-dominated society has been so persistent in institutionalizing the double standard.

But *Homo sapiens* is a peculiar creature, influenced by many things beyond its biology. Given any biological predispositions in a particular direction, however slight, we often extend these inclinations by cultural prescriptions and injunctions, sometimes even hyperextending them far beyond any reasonable scope provided by biological underpinnings. The sexual double standard may well be a "cultural hyperextenson" of this sort, an instance of human societies taking a biological molehill and exaggerating it into a mountain of male–female differences.

Having looked, although briefly, at the diversity of human mateships, what can we conclude? For one thing, it seems undeniable that human beings have evolved as mildly polygynous creatures whose "natural" mating system probably involved one man mated, when possible, to more than one woman. It is also clear than even in societies that institutionalized some form of polygyny, monogamy was nonetheless frequent, although, for men at least, this typically meant making the best of a bad situation. (Worse yet was bachelorhood.) There is also great diversity, however, in the patterns of monogamy, ranging from frequent extramarital sexuality, condoned and sometimes even encouraged by the social code, to occasional affairs, frowned upon but not taken very seriously, to rigid monogamy, jealously and violently enforced . . . although even here it seems likely that the rules of absolute sexual fidelity are often violated, in secret.

Certainly there is no evidence, either from biology, primatology, or anthropology, that monogamy is somehow "natural" or "normal" for human beings. There is, by contrast, abundant evidence that people have long been prone to have multiple sexual partners.

In a sense, however, even if human beings were more rigidly controlled by their biology, it would be absurd to claim that monogamy is unnatural or abnormal, especially since it was doubtless the way most people lived, most of the time . . . even while men strived for polygyny and women (as well as men) engaged in EPCs. This is clearest for men, if only because polygyny has often been institutionalized—and, thus, proudly displayed by the male "winners"—whereas EPCs among *Homo sapiens,* as among most living things, have been much more covert, because of the costs of disclosure. Nonetheless, male philandering would never have become part of our biological heritage if women did not permit some men, at least on occasion, to succeed in their quest for EPC. Which is to say that, whether officially polygynous or monogamous, women—perhaps no less than men—have *long* sought extramarital lovers.

Human beings are enormously flexible creatures, at least socially. They are capable of living a variety of lives depending on the demands and expectations of the society in which they live. To some extent, then, human inclinations may be able to fit whatever matrimonial pattern happens to exist in the society they happen to experience.

But, on the other hand, mild polygyny is likely so much a part of the primitive human condition that it is reflected not only in our anatomy, physiology, and development—not to mention the anthropological record—but also in our behavioral tendencies. If this is true, then the marriage bed may be a procrustean bed as well, insofar as it denies the possibility of nonexclusive sexual relationships. Deprived of both socially approved polygyny and EPCs, perhaps it is not surprising that many people, throughout history and around the globe, have chafed at lifetime monogamy and often circumvented it.

What makes human beings unusual among other mammals is not our penchant for polygyny but the fact that most people practice at least some form of monogamy. At the same time, *Homo sapiens* is quite prone to sexual jealousy, which strongly suggests that monogamy has long been unstable.

Psychiatrist Wilhelm Reich is an interesting case. Reich insisted in his work and his writings that monogamy was an unhealthy state for human beings, undermining their sexual health and stunting their emotional lives. Yet his wife reports that Reich was often insanely jealous:

> Always, in times of stress, one of Reich's very human failings came to the foreground, and that was his violent jealousy. He would always emphatically deny that he was jealous, but there is no getting away from the fact that he would accuse me of infidelity with any man who came to his mind as a possible rival, whether colleague, friend, local shopkeeper, or casual acquaintance.

Among animals, male–male competition is the centerpiece of the most consistently aggressive and often violent actions that take place within a species, including the great skull-splitting clashes of bighorn sheep and the ferocious natural battles that span the animal kingdom from whales to dungflies. Small wonder, then, that even some of the most perceptive, avowedly liberated, and otherwise civilized members of *Homo sapiens* sometimes "lose it" when it comes to sexual jealousy and male–male competition.

Even Sigmund Freud, so insightful—and sometimes fanciful—with regard to unconscious mental processes, was afflicted with jealous rages. In one of his many hundreds of letters to his fiancée, Freud noted his reaction upon finding out that she had once encouraged another suitor to express his affection for her: "When the memory of your letter to Fritz . . . comes back to me I lose all control of myself, and had I the power to destroy the whole world . . . I would do so without hesitation."

Freud later proposed that sibling rivalry developed as a result of the older child's fundamental jealousy at being displaced in the mother's affections by a younger one. His interpretation may be more appropriate if applied to sexual jealousy, which in fact is more aptly described in the following passage devoted to sibling rivalry:

> What the child grudges the unwanted intruder and rival is not only the suckling but all the other signs of maternal care. It feels that it has been dethroned, despoiled, prejudiced in its rights; it casts a jealous hatred upon the new baby and develops a grievance against the faithless mother.

In one of Shakespeare's lesser-known plays, *Cymbeline*, Posthumus is made to think that his wife has been unfaithful. When shown her ring, which convinces him of his wife's adultery (although actually she has been altogether faithful), Posthumus cries out his jealous anguish:

> It is a basilisk unto mine eye,
> Kills me to look on't. Let there be no honor
> Where there is beauty; truth, where semblance, love
> Where there's another man. The vows of women
> Of no more bondage be to where they are made
> Than they are to their virtues, which is nothing.
> O, above measure false!

As his overheated mind dwells on the imagined act, the husband's words become more and more specific: "No, he hath enjoyed her." And later: "She has been colted by him." Eventually, his wife's imagined single act of adultery becomes equivalent to innumerable such events: "Spare your arithmetic; never count the turns. Once, and a million!"

Jealousy is one of those things that appears to make no sense, yet afflicts us anyway. Indeed, some of our most deep-seated traditional teachings are ambivalent—if not downright contradictory—as regards jealousy. The Old Testament spoke approvingly of the emotion—at least in High Places: "I the Lord your God, am a jealous God" (Deuteronomy 5:9). And yet St. Paul cautioned: "Love is not jealous or boastful" (I Corinthians 13:4).

By and large, jealousy is not one of our more admired emotions. Shakespeare's Iago called it "the green-eyed monster which doth mock the meat it feeds on." John Dryden described jealousy as "the jaundice of the soul," and in *Paradise Lost*, Milton referred to it as "the injured lover's hell." The English poet Robert Herrick derided jealousy as "the canker of the heart." And yet the Roman poet Lucian opined that "When a man is not jealous he is not really in love." Insofar as there is a possessive quality to love—and there may always be, to some extent—love rarely exists without jealousy. Would an absence of jealousy simply be an absence of possessiveness—or an absence of caring?

According to anthropologist Ruth Benedict, there is a connection between jealousy and passionate intensity: You can't have one without the other. In her description of the Dionysian inhabitants of the island of Dobu, she recounts:

> Any meeting between man or woman is regarded as illicit, and in fact a man by convention takes advantage of any woman who does not flee from him. It is taken for granted that the very fact of her being alone is license enough. Usually a woman takes an escort, often a small child, and the chaperonage protects her from accusation as well as from supernatural dangers.

The deep-seated prudery of Dobu is familiar enough from our own cultural background, and the dourness of Dobuan character that is associated with it also accompanied the prudery of the Puritans. But there are differences. We are accustomed to associate the Puritans with a denial of passion and a lesser emphasis on sex. But this disassociation of prudery and passion is not inevitable. On Dobu, prudery coexists with prenuptial promiscuity and a high valuation of sexual passion and techniques. Men and women alike rate sexual satisfaction as highly important and make achievement of it a matter of great concern. The stock sex advice given to women entering marriage is that the way to keep a husband is to keep him as sexually exhausted as possible. There is no belittling of the physical aspects of sex. The Dobuan, therefore, is dour and prudish, consumed with jealousy, suspicion, resentment—and also sexual passion!

Adultery, jealousy, and violence often make for a lethal mix. Among the Venezuelan Zorcas, an adulterous woman isn't punished by the tribal leaders . . . provided that her husband kills the lover. The ancient Maya allowed the husband to decide whether his wife's lover should be killed. Male responses to adultery are not invariably violent, however, with or without the involvement of civil authorities. We have already seen cases in which wives are traditionally exchanged, and there have even been some societies—although not many—in which EPCs are seen as "no big deal." In certain Eskimo groups, men responded to their wives' adultery by challenging the lover to a public song contest. The Gabriellino Indians of southern California had an even less resentful and somewhat more practical solution: The offended husband could give his wife to her lover and take the other's wife as his own.

A woman is generally less likely than a man to behave violently—including, under most circumstances, in response to adultery—but she is is not necessarily any less likely to be emotionally hurt, even infuriated by a spouse's infidelity. As William Congreve put it: "Heaven has no rage like love to hatyred turned; Nor hell a fury, like a woman scorned." The "scorned" wife is the closest English equivalent to a cuckolded husband, although the former part of the Congreve phrase has never been quoted as widely as the latter.

For all the power of jealousy, however, it is noteworthy that, despite nagging doubts, most people retain substantial confidence in monogamy as an institution generally and in the monogamous inclinations of their partner in particular . . . regardless of whether either is justified. "It is the property of love," wrote Marcel Proust in *Remembrance of Things Past*, "to make us at once more distrustful and more credulous, to make us suspect the loved one, more readily than we should suspect anyone else, and be convinced more easily by her denials." Whatever the human penchant for jealousy, the desire to think well of a loved one is generally even greater. The facade of fidelity is terribly important to most people, although, to be sure, there are those who enjoy being publicly victimized in the hope of gaining spectator sympathy.

In general, however, virtually no one wants his or her spouse to be unfaithful, and virtually everyone will go to great lengths to ignore or deny evidence to the contrary. One might think that a bit of introspection would convince most people that even a much-loved and much-loving spouse is at least capable of "going outside" a marriage, any marriage, should the opportunity arise.

Behind sexual jealousy and the widespread human concern with adultery there lurks—as with other animals—concern about parentage. Such concern need not be conscious and may even be denied if made explicit; there is no reason to think, for example, that childless couples are any less prone to sexual jealousy than are "breeders." The point is that intentional childlessness is relatively new to the human experience; after all, genes for *not* reproducing would generally face a rather bleak evolutionary future! And so we have almost certainly evolved with reproduction-relevant tendencies, no less than have doves, dunnocks, or donkeys. And once again, the spotlight is especially on men, not because they are more important than women but because they are *less* important: Whereas mammalian mothering is obligatory, fathering is more problematic. Not surprisingly, for mammalian fathers in particular, their mates' EPCs pose a special problem.

Female birds are generally more prone to desert than are female mammals, at least in part because male birds are as capable as females of doing the parenting chores, whereas female mammals are to some extent tied by their breasts to their kids. At the same time, it seems likely that EPCs by their mates are more hurtful to male birds than to male mammals, since, among birds, males can do—and actually do—much of the child-rearing. (A cuckolded male mammal at least doesn't end up lactating for someone else's offspring!) We might expect, therefore, that male birds might object more strongly to female infidelity. On the other hand, human beings are unusual—perhaps unique—in the amount of parental care often provided by males. So we can expect that, among human beings, infidelity by a mate would typically be seen as a very grave offense. It is.

Considering all mammals, primates are the subgroup (technically, the "order") among which fathers are most involved with their offspring. Perhaps 40 percent of mammalian genera have some form of father–offspring interaction. At the same time, infanticide—the polar opposite of paternal aid—is also frequent among primates. This is not contradictory, however, because it is probably the substantial amount of paternal involvement that inclines nonpaternal males to kill juveniles to whom they are not related, so that their paternal care goes only to those young to whom they *are* related.

Monogamy, too, is somewhat more common among primates than among most mammals, but as already pointed out, it remains rare, and the more we know of the private lives of the generally secretive monogamous primates (or rather, those purported to be monogamous), the fewer of them there appear to be. Human beings—although, to be sure, not reliably monogamous—are more monogamous than most primates and far more so than most mammals. Maybe men would be even better fathers if women were more reliably monogamous. (Which, in turn, would require that men be more reliably monogamous, too!)

It may be a tall order. A review of 56 different human societies found that in fully 14 percent nearly all women engaged in EPCs, whereas in 44 percent, a moderate proportion did so, and in 42 percent relatively few—but still some—did so. It is revealing to compare these figures with those of their male counterparts: Nearly all men engaged in EPCs in 13 percent of societies, a moderate proportion of men did so in 56 percent, and a few—but still some—did so in 31 percent. In short, cross-cultural analysis of infidelity rates shows that females and males are remarkably similar.

The United States is no exception. According to Kinsey and colleagues, slightly more than one-fourth of adult females in the United States had engaged in EPCs. A *Cosmopolitan* magazine poll reported numbers closer to 50 percent (perhaps reflecting the readership of *Cosmopolitan*). A different survey found that 12 percent of women had been sexually unfaithful to their husbands during their first year of marriage; after 10 years of marriage, this number rose to 38 percent.

Compared with the high probability of male retribution after female infidelity, it is quite rare for male infidelity to trigger female retribution. (Female retribution, similarly, is almost unknown among animals, although interference, as we have seen, is not uncommon.) Among human beings, about 75 percent of societies permit male infidelity, whereas only about 10 percent permit female infidelity—and even in these cases, it is not guaranteed that males will actually be tolerant of such behavior. One can also predict a correlation between male suspicion of female EPCs and male neglect of, or even violence toward, that female's offspring.

Earlier, we considered the notion that multiple matings might serve several different animal species as a form of infanticide insurance; this would be a risky strategy, however, in *Homo sapiens,* whose large brain makes possible both a penchant for suspicion and an ability to put 2 and 2 together. Thus, female infidelity—if detected—could make infanticide *more* likely.

Among the Aché people of eastern Paraguay, women commonly rely upon their former lovers to provide not only food but also protection. Seventeen different Aché women were interviewed by a pair of anthropologists; among them, these women had 66 offspring, which were attributed—by their mothers—to an average of 2.1 fathers each! Interestingly, this was pretty much the optimum, because the chances of survival declined for an Aché child who had more than two or three possible fathers, apparently because no one felt sufficiently confident of fatherhood to help out.

The Aché, according to the interviewing anthropologists, recognize three different types of fatherhood: "One type refers to the man who is married to a woman when her child is born. Another type refers to the man or men with whom she has had extramarital relations just prior to or during her pregnancy. The third type refers to the man who *she* believes actually inseminated her." It seems likely that Aché men, for their part, make parallel distinctions, at least concerning children they might have fathered versus those they definitely did not. This has not as yet, however, been reported in the literature.

On the other hand, paternal care need not be driven strictly by confidence of genetic relatedness. The men's "quality" seems likely to be relevant as well, although, at this point, we cannot predict how the correlation will go. For example, males in general and men in particular who are in poor condition might contribute relatively little to the care of their offspring, for no other reason than that they are less able to do so, whereas males in good condition might contribute more, simply because they can. Low-quality, less-desirable men might also be cuckolded more, as has been documented for numerous animals. If both are true, then such men would both have a low chance of being related to their offspring and display a low level of paternal involvement—and yet the two would not be causally connected.

It is also possible, however, that a male in poor condition would invest *more* in his offspring; if he has a low chance of surviving, he might therefore place all his bets on his present family. For their part, women—like females in other species—might well be prepared to invest more in offspring fathered by attractive males. If those offspring were generated via EPCs, and if the high-quality mate of a heavily investing female is himself inclined to invest less (because she is investing more), then there would once again be a correlation between low probability of paternity and low paternal investment—although, once again, one not mediated by confidence of genetic relationship!

Historically, it is at least possible that the combination of monogamy plus adultery was as important—or more so—than polygyny. In general, adultery is more common than polygyny, at least in hunter-gatherer societies, simply because it is very difficult for a man to acquire more than one wife—and to keep her. This might also help explain our human penchant for a high frequency of sexual intercourse, as follows.

Among those species that copulate very often—lions, bonobo chimpanzees, acorn woodpeckers—it seems that the threat of sperm competition is the driving force. But these animals do not even make a pretense of monogamy. A better model for us humans might well be those socially monogamous species—notably certain birds, such as the white ibis—in which males as well as females make a significant investment in care of the young and which are driven by ecological considerations into living in high-density colonies. Socially monogamous nonhuman primates generally live such isolated lives that the threat of EPCs is greatly reduced . . . even if the predilection still persists. But as noted, despite the fact that human beings are biologically inclined to polygyny, most people end up being socially monogamous and also—like our fathered friends—living in dense "colonies" known as towns, villages, or cities.

Under these circumstances, with a premium placed on biparental care but also a substantial risk of cuckoldry, it makes sense that *Homo sapiens* is a pretty sexy creature, inclined to engage in lots of sexual intercourse. Admittedly, it is unromantic to see much of human lovemaking as driven by the threat of adultery and, hence, the need to prove one's worth, not to mention the even more "mechanical" payoff that comes from prevailing in sperm competition. Many will reply that we make love a lot (by most animal standards) because we love each other a lot, or because it feels good, or simply because we like to do so. But *why* is sexual intercourse so closely connected to love in human beings? *Why* does it feel so good? *Why* do we like to do it, even when we aren't interested in reproducing . . . sometimes especially when this is the case?

Although there are exceptions (some of which we have reviewed), among most species of monogamous mammals, sexual behavior is neither especially frequent nor especially fervent. For the great majority of socially monogamous mammals, sex takes a back seat to resting together, mutual grooming, and simply "hanging out." In most species, when the pair-bond is well established, relatively little energy is expended on sex or obvious social interactions of any sort. Yet sexual behavior is prominent among *Homo sapiens,* often identified as an important component of love and, thus, monogamy. Maybe our unusual preoccupation with *sexual* love has developed because, unlike the great majority of monogamous mammals, which live rather isolated, hermit-like lives, for which the risk of EPCs is very low, we are highly social and, thus, sorely tempted to stray. Seeking to maintain a degree of monogamy despite living so close to one another, perhaps human beings have added lots of sex both as a way of reconfirming and, if need be, reestablishing the pair bond, while also meeting the demands of sperm competition by providing men with sufficient confidence of genetic relatedness for them to invest in their mate's offspring.

And so we come to sperm competition, a difficult topic in many ways. It is difficult to study, difficult to arrive at firm conclusions about, and emotionally difficult because it challenges some of the deepest and most anxiety-ridden undercurrents of our emotional lives. Especially the lives of men. (As we shall see, women may or may not set up conditions for sperm competition; if they do, then men—knowingly or not—have to participate.)

The big question is this: Is sperm competition a significant factor for human beings? Recall that sperm competition takes place when the sperm from more than one male compete to fertilize the eggs of a female. For sperm competition to be important among human beings, during the course of human evolution—possibly continuing into modern times as well— women must have frequently had successive episodes of sexual intercourse with more than one man during a brief time span; that is, while they were fertile. Strictly monogamous women could not promote sperm competition. Women who are polyandrous, promiscuous, prostitutes, rape victims, or socially monogamous but also prone to EPCs could.

The most ardent advocates of the importance of human sperm competition are two British biologists, Robin Baker and Mark Bellis. Their work is controversial, accused by critics of being inadequately supported by the data and sometimes verging on shallow sensationalism, while lauded by supporters as being bold, innovative, and path-breaking. The jury is still out, while the evidence is just starting to dribble in.

It appears that the average mating interval for moderately young, healthy heterosexual couples is three days. Whether coincidental or not, such a frequency of IPCs—intra-pair copulations—maintains an almost continuous supply of sperm within the female's reproductive tract. EPCs, by definition, tend to lead to sperm competition (except, of course, when the female has not been having sexual intercourse with her spouse and she has only one EPC partner). Not surprisingly, it is very difficult to say how frequent human EPCs are and, more specifically, how frequently a woman will mix sperm from more than one man. To carry around a mix of sperm, a woman would have to have sex with two different men within about a five-day interval.

IPCs are pretty much evenly divided throughout a woman's reproductive cycle, if anything somewhat more frequent during the postovulation phase, when fertility is substantially reduced. By contrast, Baker and Bellis report that EPCs are actually more frequent when women are most fertile! According to the two researchers, "at some time in their lives the majority of males in western societies place their sperm in competition with sperm from another male and the majority of females contain live sperm from two or more different males." They estimate that in Great Britain 4 to 12 percent of children are conceived by "sperm that has prevailed in competition with sperm from another male." This is consistent with standard estimates of "paternal discrepancy" among human beings generally: about 10 percent, which, if accurate, is enough to bespeak genuine sperm competition.

Mixed fatherhood is most dramatic in certain cases of nonidentical twins, for example, when one infant turns out to be Caucasian and one Asian, and so forth. The most famous "twins," Castor and Pollux, were said to have had two fathers, one being Zeus and the other, their mother's mortal husband. In a survey of nearly 4,000 sexually experienced women (having had at least 500 copulations), 1 in 200 claimed to have had sexual intercourse with two different men within 30 minutes of each other on at least one occasion; within 24 hours, the number jumped to nearly 30 percent. If these data are reliable, they, too, suggest plenty of opportunity for sperm competition.

Most insects and birds have a "last male advantage," which means that the last male to mate with a female is likely to fertilize the lion's share of her eggs. In the case of mammals, however, the situation is much less clear. It is completely unknown whether, among primates in general or human beings in particular, the first male, the last, or anyone in between has any advantage. The probability of genetic fatherhood among human beings may simply be determined by having the most sperm in play, the reproductive equivalent of the old military adage of being "firstest with the mostest." Regular and frequent IPCs can be seen, therefore, as a means of "topping off" a woman's reproductive tract, replacing sperm likely to have died or become disabled since the last coitus and thereby maintaining a more or less constant and optimal level of sperm population. This would be adaptive under any circumstances, simply to maximize the chances of fertilization. But it would be especially appropriate in an environment of sperm competition; that is, when the woman might copulate with other men as well.

Baker and Bellis point out that both men and women have numerous ways—nearly all of them unconscious—of influencing the outcome of sperm competition. Women first.

Women are not merely passive vessels within which men and their sperm carry out a series of competitive events. They can, and do, exert substantial choice, evaluating men by numerous criteria—emotional, physical, intellectual, financial—in a search for appropriate qualities as a parent, protector, friend, colleague, lover. And like other living things, there is no reason why women could not find themselves socially mated to one male but inclined to engage in EPCs with another. It is noteworthy that many female primates give loud calls during copulation. Among free-living baboons, these calls are most frequent when their sexual swelling is most intense: Females called in 97 percent of copulations. Such calls are longest when ejaculation takes place (which, for baboons, is only about one-third of the time). This suggests that females call to encourage other males and, thus, to promote sperm competition among them.

Human women are no less likely to be concerned about the overall quality of the males who fertilize their eggs, although there is no evidence for anything quite as immediate and explicit as the baboon system of mobilizing competition. The internal reproductive tract of women produces its own barriers as well, including antisperm antibodies that can interfere with fertilization by immobilizing or even destroying sperm and impairing their ability to penetrate the egg, while other antibodies act against the egg's membrane, preventing early egg cleavage and development. A key point is that these antibodies do not necessarily reduce absolute fertility; rather, they diminish the fertility of particular male–female pairs. A woman caught in this biochemical fertility trap may enhance her reproductive success—whether she consciously realizes it or not—by seeking a different reproductive partner while perhaps still retaining her social, marital partner.

Aside from the array of behavioral strategies of which women are capable, it seems likely that they also manipulate sperm directly . . . although, again, not consciously. Especially prominent among these sperm manipulations is "flowback." Up to one-third of the seminal fluid deposited within the vagina leaks out within a few minutes of intercourse. Semen is also discharged when urinating, at which time it is expelled with substantial force, as compared to dribbling out after coitus when a woman stands up—or even if she remains lying down. About 12 percent of the time, this flowback results in the expulsion of essentially all sperm deposited inside a woman's reproductive tract. So, at this level alone, women are capable of exercising substantial control over sperm (recall the Atalanta solution, discussed earlier for other species).

Baker and Bellis surveyed women who were either married or otherwise involved in a serious one-to-one heterosexual relationship and evidently obtained the cooperation of many, who were remarkably open about some of the most private aspects of their lives. They report that women are especially likely to engage in EPCs when they are fertile, suggesting an unconscious strategy of choosing more desirable males, via EPCs, as potential fathers for their offspring . . . even if such a strategy is intentionally thwarted by use of birth control. Baker and Bellis also somehow convinced a number of women to capture flowback after both EPCs and IPCs; the results show a *lower* level of sperm retention associated with sexual intercourse with their main partner. In other words, not only are women more likely to engage in EPCs when they are more fertile, but they also retain more semen after such encounters. (According to Baker and Bellis, women achieve higher sperm retention during EPCs by reducing their frequency of noncopulatory orgasms via masturbation; in other words, they claim that contractions during female orgasm actually push out semen and that by masturbating less and thus having fewer orgasms, women end up retaining more EPC sperm.)

It has long been known that insects and birds have sperm storage organs, whereas mammals, it was claimed, lack anything comparable. Nonetheless, the argument has been made that sperm are stored by the millions in so-called cervical crypts, tiny cavities lining a woman's cervix. From here, they could be released over a period of hours, even days, after intercourse, with peak release over a period of 2 to 24 hours. This is important for our purposes because the ability of women to store sperm from successive copulations sets the stage for sperm competition among successive males.

There is also debate about how much time can elapse between two successive copulations with different men in order for sperm competition to occur. A key question is how long after ejaculation sperm can remain viable and, therefore, capable of duking it out for the privilege of fertilizing a woman's egg. Minimum estimates are two to three days; the maximum seems to be seven to ten. A reasonable number may be five to six. What this means is that if a woman has sex with someone within five or six days of having had sex with someone else, the sperm from these two men could be in direct competition.

How does one "win"? For women, it means making the best choice; that is, having the opportunity of choosing among more than one sperm provider (hence, copulating with more than one man), as well as being able to make a "good" choice among the available sperm, perhaps by setting up a competitive situation. For men, winning means having one's sperm succeed. We turn to them next.

Among men, sperm strategies might include simply making lots of the little pollywogs, assuming that fertilization is a kind of lottery. Buy more tickets and you're more likely to hit the lottery jackpot. Make more sperm, and deliver them to the right place, and you're more likely to hit the fertilization jackpot. Even in this case, however, different tactics present themselves. For example, biologists have asked, "What is the optimum strategy for males, in terms of partitioning their sperm in IPCs versus EPCs?" Geoffrey Parker concluded from a detailed mathematical model that already-mated males should generally ejaculate more sperm during EPCs than IPCs, assuming that such males are normally able to maintain an adequate sperm level within their in-pair mate. The only exception would be when a male has determined that his female has engaged in one or more EPCs, in which case he should increase his sperm numbers during IPCs. It remains to be seen whether such adjustments actually occur.

But what if fertilization is less a lottery than a race? Then it would be important to make sperm that move quickly. Or maybe it's a war, in which sperm from different men literally do battle with each other.

Baker and Bellis have accordingly proposed their "kamikaze sperm hypothesis." It suggests that only a very small proportion of human sperm are intended to function as "egg-getters." All sperm, they claim, are not created equal, nor are they created to do the same thing; namely, to fertilize eggs. Thus, sperm are not homogeneous little packages of DNA, each pruned down to the minimum size needed to achieve fertilization. The smallest sperm in a single human ejaculate, for example, can have merely 14 percent the volume of the largest. There is more variation within a single human ejaculate than in the mean sizes of sperm from all the different primates. In short, a normal human male produces a remarkable range of different types of sperm: amorphous, pin-sized, weirdly shaped, double-headed ("bicephalous"), crook-necked, double-tailed, short-tailed, coiled-tailed. Previously, this sperm diversity was thought to be simple pathology: Easily 30 percent of human sperm were acknowledged to be defective in some way. (Indeed, this high rate of "bad sperm" has long been thought to be one of the reasons why men make so many of them.) But if selection had simply acted on males to produce egg-getters, why should so many be defective, slow, lame, and seemingly deformed? In most other contexts—even those that seem comparatively more trivial—natural selection does a much more efficient job.

Baker and Bellis argue that semen should be seen as another human organ, comparable to the liver, the kidneys, or—more to the point—the immune system. As such, it is composed of many different kinds of highly specialized cells, all of which together contribute to getting an important job accomplished, one of which is doing battle on behalf of the rest of the body. Among rats, sperm that form copulatory plugs are those with smaller heads, which are more likely to become decapitated. Rat sperm from one copulation thus get in the way of sperm from the next. It is not impossible—although as yet unproved—that something similar happens among human beings, too. We don't produce copulatory plugs, but the sperm from one man could nonetheless interfere with those of another; indeed, there could be an evolutionary payoff to men whose sperm are especially nasty to anyone else's.

Baker and Bellis maintain that the majority of human sperm are, in effect, kamikazes, on a suicide mission whose goal is simply to block the sperm of other males. In addition to these "blockers"—sperm with coiled and kinky tails—there are others that go out on "search-and-destroy" missions. These seem especially prone to chemical warfare, via specialized structures, known as acrosomes, that adorn their tips.

Baker and Bellis claim that when sperm from two different men are mixed, many are disabled or killed, whereas this does not happen when ejaculate from one man is separated, then recombined. This suggests that something like an immune response—a form of chemical competition—goes on between sperm produced by different men. If so, it is possible that older sperm become blockers, since this requires less energy and vitality, while younger ones are designated to carry the ball . . . and, if they are especially fortunate, to "score." (Indeed, whereas newly minted sperm have a low proportion of coiled tails, the older the sperm, the greater the proportion of coiling.)

Further evidence of male adaptations to sperm competition comes from assessing the detailed makeup of the male ejaculate. Human ejaculation takes place in a series of three to nine spurts, closely linked together in time. Herculean efforts by researchers (and their subjects) have enabled examination of so-called split ejaculates, obtained by capturing a few squirts of semen from various stages of ejaculation. The results: Early and late spurts are different. The final pulses actually contain a spermicidal substance, which might well be instrumental in ambushing those of the next male likely to ejaculate in the same female! At the same time, chemicals present in the first half of the human ejaculate contribute to some protection of sperm against those chemicals present in the second half . . . and thus, also, possibly, against any chemicals deposited by the final spurts of any preceding male.

It is well known that men produce larger ejaculations when their sex lives have been interrupted, and then resumed. By itself, this appears to be a simple physical consequence of whether or not seminal fluid has had a chance to accumulate over time; with regular emptying, the volume of any one ejaculation is necessarily less. It is not surprising, therefore, that Baker and Bellis found that when a man spends time away from his female partner, he produces more sperm per ejaculation once sexual relations are resumed. More remarkable is their discovery—by analyzing the condom contents of some extraordinarily cooperative subjects—that sperm concentration is higher when such men actually engage in sexual intercourse than when they masturbate. (A sperm-competition perspective, incidentally, also suggests that masturbation may be a way of ensuring that the sperm available to be ejaculated in sexual intercourse have a relatively long "shelf life"—by getting rid of older sperm, thus ensuring that what's left is fresh. The likelihood is that younger sperm are better able to compete, especially more able to penetrate the cervical mucus. Baker and Bellis also suggest that older sperm may serve a secondary role, as blockers, guarders, kamikazes, troublemakers, troubleshooters, and so forth.)

In addition, during copulation, the amount of sperm transferred seems to be further adjusted according to the risk of sperm competition, especially how long it has been since the last copulation with the same woman, and even how much time the two have spent together during the previous few days. As Baker and Bellis put it, "Males may not look very sophisticated in the moments leading up to and during ejaculation but . . . some very sophisticated adjustments are taking place."

There is yet more.

Consistent with their courageous—or foolhardy—willingness to probe some of *Homo sapiens'* most intimate secrets, Baker and Bellis have also looked at the human penis. Although not remarkable by mammalian standards generally, the human penis is the largest among all primates, and Baker and Bellis suggest that its size and shape may also have been sculpted by sperm competition. They point out that in less than a minute after ejaculation, semen forms a soft, spongy coagulate, a mass that is vulnerable to being removed by a subsequent sexual partner, assuming that he encounters the woman quickly enough and that he is adequately equipped to do the job.

There is nothing unique, by the way, about penises being designed to compete with other males. In Chapter 2, we encountered the remarkable, competition-oriented penises of damselflies and sharks. Human beings clearly have nothing directly comparable; indeed, the penis of *Homo sapiens* is notably unadorned as such organs go. Nonetheless, the long-standing male obsession with penis size and shape may in itself suggest an accurate, if unconscious, recognition that this organ may actually be about as consequential as the most ardent Freudian or locker-room comparator has generally assumed.

For Baker and Bellis, the length of the penis as well as its enlarged bulbous tip suggest its role as a "suction piston" for sperm removal, a kind of natural "plumber's helper," designed to break up and possibly even remove coagulated sperm deposited by a previous male. This effect would be enhanced by the often vigorous thrusting that characterizes sexual intercourse and ejaculation among humans . . . and which is surprisingly difficult to explain or justify otherwise.

Let's further develop this admittedly homey metaphor: A plumber's helper is used to push blockage farther down the system, which, in our context, would seem counterproductive, actually facilitating fertilization by the *earlier* male's sperm. Except that the female reproductive tract is essentially a dead end, so the "suction piston" hypothesis might have some validity after all. Another potential problem: If sperm competition has been so important in producing men's anatomy, especially the seemingly oversized penis, why do men have proportionately smaller testicles than chimps? Possible answer: Maybe sperm competition is even more important for chimps, since however important EPCs may have been for humans, they are even more prominent among *Pan troglodytes*. But then, why don't chimps also have large, competitive penises? No one knows. (Nor, to our knowledge, has anyone asked.)

For obvious reasons, vagina and penis are closely matched in all species, not unlike a lock and key. So human penis size may have been largely determined by the size of the vagina, rather than by the dictates of sperm competition among males. Vagina size, in turn, seems to have been determined by the size of the infant—specifically, its head diameter—that must pass through in childbirth. As human head size increased during human evolution to favor large brains, so, presumably, did the size of the vagina. And this may in turn have generated pressure for the evolution of larger, more competitive penises as well.

Robin Baker and Mark Bellis are fervent proponents of the role of sperm competition among human beings. And they may well be onto something. But as the great nineteenth-century statesman Talleyrand suggested for his diplomats: "Pas trop de zele" ("Not too much zeal"). In science, too, zeal can be troublesome, especially when it leads to the temptation to sensationalize, to overgeneralize, to ignore contrary evidence. Even though Baker and Bellis's arguments are so convivial to the thrust of the present book that we have presented them at length, it must be noted that nearly all of their conjectures are still that, far from proven. The dirty little secret of science is that it is not done by omniscient deities or computer-driven robots but by scientists, who are fallible human beings. Although we seek ultimately to unravel genuine external truths about the natural world, not simply to validate our own preconceptions, one of those truths is that we are readily seduced by our own ideas and just as reluctant to give up on them—even in the face of contrary evidence—as anyone else.

Thus, it may be that sperm competition itself is less important than this book, or Baker and Bellis, has suggested. A study of chimpanzees in which 1,137 copulations were observed found that more than 70 percent involved multiple matings; that is, females mating with more than one male. Only 2 percent took place during isolated one male–one female consortships. So far, so "good," at least insofar as the case for sperm competition is concerned. Yet most conceptions occur during these essentially monogamous consortships (which take place during maximum tumescence and, thus, peak fertility). Such findings urge caution: Even behavior such as promiscuous chimpanzee sexuality, which screams "sperm competition," might actually involve less than meets the ear, no matter how favorably attuned.

Human beings, moreover, are less prone to EPCs than are chimps. As evidence, human sperm concentration diminishes more rapidly with repeated copulations, suggesting that men are less adapted than male chimps to competing with the sperm of other men. As with other species, human sperm are cheap, but not free. In one intriguing experiment, men engaged in a "10-day depletion experience," averaging 2.4 ejaculations per day. Afterward, their sperm output remained below their earlier, predepletion levels for more than five months! By contrast, male chimps—who have to deal with polyandrous, EPC-inclined females—can ejaculate every hour for five hours, after which their sperm count is only halved, with very rapid recovery. So whatever the importance of sperm competition for *Homo sapiens*, it is likely not as pronounced as it could be.

Others, notably biologist Alexander Harcourt, have criticized specific aspects of Baker and Bellis's work more directly, pointing out that mammals are unlikely to produce designated, nonfertilizing sperm for several reasons:

given a high natural loss, males may well be unable to afford production of sperm that are guaranteed not to be potential fertilizers and whose supposed aggressive/defensive activities may never even be called for. Moreover, secretions from the accessory gland of males seem sufficient by themselves to coagulate semen and generate copulatory plugs, at least in other animals; a male who used these secretions for such a purpose and continued to produce fertilizing sperm instead of diluting his ejaculate with kamikazes would be at an evolutionary advantage.

Most impressive, however, is this finding: Males of polyandrous species are under more intense sperm competition than in monandrous species (whether monogamous or polygynous) in which females mate with only one male. And yet a review of research findings shows that polyandrous species do not produce a greater number, or even a higher proportion, of deformed (and, presumably, nonfertilizing) sperm. Nor do they produce more slow-swimmers, which might be expected if such kamikazes were specialized to stay behind and duke it out with a competitor's sperm. Harcourt concludes that sperm competition—at least in mammals—occurs via what ecologists call "scramble competition," in which contestants struggle individually toward a goal, irrespective of their fellows, as opposed to "contest competition," in which individual contestants would seek to best their fellows *mano a mano* (rather, spermo a spermo).

The male penchant for producing vast numbers of sperm may therefore exist because fertilization is a simple "raffle," rather than a direct competitive struggle. (Even a raffle would still involve sperm competition, but one in which the contestants compete by buying as many tickets as possible rather than by tearing up each other's entries.) Or maybe males make lots of sperm simply because, considering their very high mortality—even *without* sperm competition—it behooves males to make lots of little wigglers if fertilization is to occur at all. After all, the low pH of the vaginal environment is as hard on human sperm as it is on other vertebrates, phagocytes roam through every woman's reproductive tract, many sperm end up being helplessly absorbed into the uterine wall, and they do have a long way to swim, not to mention the fact that fully one-half can be expected to swim up the wrong side of the uterus, ending up perhaps at the right (wrong) fallopian tube when a fertile egg is waiting at the left, and vice versa. In short, the importance of sperm competition may simply be overblown.

But we doubt it.

Any concept comes across as all the more powerful when it sheds new potential light on old facts. For example, take pornography. It, too, could be related to sperm competition, as follows. Earlier, we mentioned the case of Grevy's zebras, in which stallions adjust their sexual performance depending on the predilections of the mares, mating more—in a sense, being more sexually aroused—when associating with females who are prone to a higher level of sexual activity. A similar pattern is found in bighorn sheep: Dominant males mate females immediately after subordinate males have done so. And it is interesting to note that male great apes, too, mate more often when their females are polyandrous. Often they are stimulated by any indications of sexual intercourse, a connection that makes sense given the presumed payoff of introducing one's own sperm to compete with those of a possible competitor.

There is little doubt that, in our own species, men are agitated and often infuriated by indications of their mates' EPCs. It is unclear whether they are sexually aroused as well, although anecdotal evidence suggests that this is not uncommon. Where does pornography fit in? Part of the excitement of pornography—especially for men—may be that it conveys the basic message that "sexual activity is going on nearby," which in turn is converted (among males especially) into competitive sexual arousal. On a primitive, biological level, pornographic images may activate the same system that prehistorically enabled men to respond to the dictates of sperm competition: If "your" woman may recently have been having sex with someone else, you would be well advised to have sex with her, too, and right away!

There is recent evidence that sperm competition is pronounced in both human beings and chimpanzees, compared to the situation in gorillas, for example. (This makes sense, since unlike chimps or human beings, gorillas live in rigidly controlled polygynous harems, in which a given female mates only with the dominant silverback male; as a result, male gorillas need to concern themselves with whatever it takes to obtain and keep control of a harem—that is, with competition at the level of bodies rather than sperm.) Researchers at the University of Chicago have found that three different genes controlling sperm function have been evolving at an especially rapid rate in human beings and also in chimps, suggesting that these two species have been experiencing substantial sperm competition, which in turn has been leading to rapid evolutionary change. By contrast, the same system has been evolving more slowly in gorillas, which makes sense given that a male gorilla pretty much monopolizes sexual access to his females, whereas this is emphatically not the case for chimps or—evidently—human beings.

But, in the interest of full disclosure, these findings could mean something very different. For example, women (like other mammals) have evolved various defenses to protect their uteruses from bacterial and other infection, and these chemical defenses—such as low pH—turn out to be harmful to sperm. So rapid evolutionary change among chimp and human sperm might not indicate competition among sperm at all, but rather the need for sperm to evolve defenses against the uterus's efforts to further its own self-defense! The question would then become whether chimp and human females are evolving reproductive tract self-defense more rapidly than are gorilla females and, if so, why. And one answer might be that insofar as the former are more likely to copulate with multiple males, they are exposed to more risks, thereby necessitating more defenses. Everything, it seems, is connected!

"The majority of women (happily for them) are not much troubled with sexual feelings of any kind." So wrote a prominent nineteenth-century physician in an influential medical textbook of the Victorian era. Happily for us all, he was wrong. Female sexual feelings, although somewhat different from those of men, are no less vigorous. Female sexuality is, however, less "up front" than its male counterpart. "We and our fantasies are the fruit of evolution," writes Natalie Angier in her esteemed book, *Woman*. "And we are waiting to be known."

For starters, there are two great mysteries of female sexuality: concealed ovulation and orgasm. The question, in both cases, is: Why? Why, for instance, are human beings so secretive about when they ovulate? We send all sorts of signals indicating our internal state: blushing when embarrassed, crying when sad, even—in some cases—conveying information we might just as soon keep private, such as the state of our gastrointestinal system when we belch or fart. By contrast, the exact (or even approximate) time of ovulation is a closely guarded female secret. There must be a reason.

Similarly, why do women have orgasms? Men's orgasm—a highly pleasurable sensation associated with ejaculation—makes obvious sense, but women have no comparable need for such a pronounced reaction to sexual stimulation. Here again, there must be a reason.

In the case of concealed ovulation, let's first note that human beings, although unusual in this respect, are not the only species to be so secretive. Others include black-and-white colobus monkeys of the African rain forest, pigmy marmosets and lion tamarins of South and Central America, blue monkeys, orangutans, woolly and spider monkeys, vervets and langurs. Nonetheless, an explanation is needed, given that our closest relatives, the chimpanzees, make a dramatic public show of their reproductive state. Human beings, for all their vaunted self-knowledge, are remarkably ignorant about their own ovulation, whereas if we were chimps, we'd have no doubt: Even a casual zoo visitor notices the vastly enlarged and gaudy pink rear end of a fertile, female chimp, whereas a woman must use an accurate thermometer or careful analysis of cervical mucus to obtain comparable information about her own body.

Why such secrecy? There are several possibilities, all of them involving EPCs. Thus, evidence is now accumulating that human females are more likely to engage in extra-pair copulations at the midpoint of their menstrual cycle, precisely when they are most fertile. This does not necessarily bespeak a conscious desire to reproduce as a result of these EPCs. It may simply be that women feel somewhat more sexually inclined when they are at mid-cycle. But this does not preclude our asking why such feelings are especially likely to involve an EPC rather than an IPC.

One might expect women to be, if anything, somewhat reluctant to have affairs when they are most fertile, insofar as they are aware of possibly getting pregnant. Counterbalancing this, however, might be an even greater reluctance to have sex—especially in a new and exciting relationship—when they are most *in*fertile; that is, when menstruating. So, avoidance of "messy sex" might lead to a tendency for women to have extramarital sexual encounters when they are *more* likely to conceive . . . even if they don't plan it that way and might, in fact, prefer it otherwise.

Robin Baker and Mark Bellis argue that women have a fully evolved unconscious preference for EPCs precisely when their fertility is highest, for much the same reasons that other species often engage in similar activities: to be inseminated by the best males and to encourage sperm competition among sexual partners. They point to the suggestive finding that women having a primary sexual partner are likely to walk farther during their monthly time of peak fertility, whereas females without a primary sexual partner show a tendency for reduced midcycle walking. This is consistent with the notion that paired women are more restless at midcycle, as a result of which they are more likely to encounter one or more new partners, whereas unpaired women are inclined to avoid midcycle contacts. Of course, the fact that something "is consistent" with a particular hypothesis is very different from saying that it "proves" anything.

In any event, there is increasingly strong evidence that women feel sexiest when they are at midcycle; that is, when they are fertile. For example, levels of estradiol (a female hormone) have been found to correlate with the kind of clothing women wear at a nightclub: Women at peak fertility wear tighter clothing, and less of it. They expose more skin than women who are not fertile.

Concealed ovulation is generally most common in primates among whom females have multiple sexual partners. A review of the scientific literature concluded that concealed ovulation evolved at most once among monogamous species, but between 8 and 11 times in cases of non-monogamy. Why? Probably because it permits females to obtain additional matings. After all, if a socially monogamous female clearly advertised her fertility, "her" male would guard her with particular care during that brief time . . . and since the rest of the time she would be infertile, her reproductive options would be limited to her social mate. (Another possibility cannot be excluded, however: that concealed ovulation led to monogamy, instead of vice versa. Maybe males were more likely to affiliate with a specific female if her time of maximum fertility was *not* identified, as a way of ensuring at least some chance of fertilizing her, assuming they were was willing to devote themselves full time to mate-guarding.)

Either way, the likelihood is that once ovulation is hidden, male guarding is less intense, if only because it has to be spread over a female's entire cycle instead of being concentrated during a few hours or days. The result appears to be that by concealing their ovulation, females grant themselves enhanced opportunity to mate with more than one male. As already described, they may also be granting themselves—or, rather, their offspring—enhanced opportunity of obtaining assistance from one or more would-be fathers. Or, at least, reduced risk of infanticide.

In at least one primate species, females clearly use sex as an inducement for males to act paternally. A field study has found that female saddle-backed tamarins (small primates of the New World rain forests), if given the opportunity, will mate with more than one adult male. Mrs. Tamarin will then give birth to twins, and a different male will proceed to assist her in caring for each infant. By spreading the breeding and giving each male an interest in her babies, the tamarin mother-to-be appears to spread the subsequent child-care duties among willing males.

For most females, genetic partners are easy to come by. Behavioral partners are a different story. Males are typically more than willing to contribute some squirts of sperm in return for a chance at reproductive success. Harder to come by are males willing to be behavioral fathers, not just genetic beneficiaries. The optimum female strategy—as in many other species—would be to get the best of each: Mate with genetically promising males and gain other benefits from wealthy, paternally inclined individuals. If all this can obtained from just one individual, so much the better. If not, then a bit of deception may be worthwhile. And concealed ovulation seems to be a useful ticket, offering the opportunity for females to present the appearances of monogamy (and, hence, to get paternal investment from their in-pair mates) while also gaining genetic benefits from out-of-pair copulations.

Of course, concealed ovulation could also be a device permitting women to engage in sperm competition while deceiving each man into thinking that he is the father. Thus, when ovulation is hidden, so is the identity of the father. The payoff, for the mother, is that she can obtain sperm from more than one male, then "choose" which to use to fertilize her egg(s). This has been called "cryptic polyandry," and, as noted, it is all the rage among birds.

Female blue tits paired to high-quality males—quality measured by anatomical traits such as longer tarsal bones, which correlate with higher survivorship and the likelihood of producing more successful young—remain sexually faithful. And females paired with low-quality males? They are inclined to be socially monogamous, but they also visit the territories of the more desirable males—and mate with them, thereby getting assistance from their in-pair mates but genes from among the best available specimens.

Another thing. Along with concealed ovulation come restraints on estrus. A woman, even at midcycle, is nothing like a bitch in heat. (If she were, then her ovulation wouldn't be concealed!) Concealed ovulation thus offers a new perspective on "reproductive choice," referring not only to a woman's choosing whether, when, or how to terminate her pregnancy, but also whether, when, and with whom to initiate it.

According to primatologist Sarah Hrdy, a female langur monkey "exhibits no visible sign when she is in estrus other than to present to a male and to shudder her head." When she encounters strange males, a female langur has the capacity to shift from cyclical receptivity (that is, a spontaneous bout of heat every 28 days) to a state of semicontinuous receptivity that can last for weeks. This gives females the opportunity to achieve sexual relationships with males other than the harem-keeper, at their own choosing, rather than being captive of automatic estrous cycles—for example, if a new and enticing male joins the troop, if a female happens to leave her troop to travel temporarily with an all-male band, or if, in Hrdy's words, "a female for reasons unknown to any one, simply takes a shine to the resident male of a neighboring troop." Other primates have a similar capacity, including several different species of guenons, vervets, and gelada baboons.

Why should women be any less endowed and, thereby, empowered? With their ovulation concealed and their sexual motivation under substantial cognitive control, women can select their reproductive partners far more easily than if they were victims of their raging hormones . . . as is the case, by contrast, for men!

Orgasm is another story, as complex and unresolved as concealed ovulation. Some, including anthropologist Donald Symons, have suggested that the female orgasm is biologically meaningless, an irrelevant but unavoidable side effect of the (clearly adaptive) male orgasm. If so, it would be analogous to nipples in male mammals—a tag-along trait, as discussed at the beginning of Chapter 3. Others, like the writer Desmond Morris, have proposed that orgasm makes fertilization more likely by encouraging a woman to remain horizontal, thereby making it easier for sperm to swim to their goal. The actual effect of female orgasm on fertilization is complex. Certainly, a woman's climax is not needed for conception to occur. Moreover, orgasm actually increases the volume of flowback, in part because it results in increased quantities of cervical mucus. There is also evidence, paradoxically, that too many sperm *reduce* the probability of conception, perhaps because of multiple fertilizations—which are spontaneously aborted—or because of the deleterious effect of chemicals secreted by exceptionally large numbers of sperm milling about.

Some female orgasms appear to increase the amount of sperm retention, whereas others are more likely to extrude sperm. Sexual intercourse itself, on the other hand, has a positive effect on conception, independent of the actual deposition of sperm. Thus, artificial insemination via a sperm donor is more likely to be successful if women also engage in sexual intercourse with their partner, even if he is totally sterile. The reason for this effect is unknown, but it emphasizes that the female sexual experience is somehow connected to the probability of conception, although the connection apparently is not a simple one.

For a time, it was thought that the capacity for female orgasm made human beings unique among animals. Not any more. For example, a study of 240 copulations involving 68 different heterosexual pairs of Japanese macaques noted that the females showed all the physiological and anatomical signs of orgasm in 80 cases (exactly one time in three). Neither the age of the female nor her dominance rank correlated with orgasm. On the other hand, the likelihood of orgasm was positively associated with duration of matings and with how active they were (literally, the number of pelvic thrusts by the male). When, by a bit of statistical sleight-of-hand, the researchers analyzed their results, taking these various measures of physical stimulation into account, what emerged was that female orgasm was most frequent among monkey pairs consisting of high-ranking males and low-ranking females, and least frequent among pairs of low-ranking males and high-ranking females. In short, social considerations are important.

This suggests another possible avenue. What if orgasm is a way in which a female's body rewards itself for having done something that is in its own biological interest? Not just mating, but mating with an especially good partner. The idea is that female orgasm, physiologically unnecessary for conception, is useful as an internal signal by which a woman's body rewards her brain for a job well done. Something similar may apply to male orgasm, too. After all, it is not strictly necessary for the pleasurable sensation of ejaculation to exceed that of, say, urinating . . . and yet it clearly does. Perhaps the intensity of orgasm says, in effect, "This is not merely a good thing to do, it is a *very* good thing!"

For men, and males generally, sexual intercourse is a biological plus (especially before the era of AIDS). Hence, a reliable, general-purpose reinforcing mechanism such as orgasm seems appropriate. For women, however, and females generally, sex is easily obtainable, but good sex—that is, sex with the right male—is harder to come by. Hence, maybe female orgasm is a way for women to confirm that their current sexual partner is especially suitable. Not necessarily as a long-term mate, mind you, just as a mate. Perhaps an ongoing EPC partner. By a process fore-shadowed by the "sexy son hypothesis" among animals, maybe female orgasm is how a woman's body italicizes that her current sexual partner—demonstrated to be capable of providing substantial sexual satisfaction—might well produce offspring capable of being similarly gratifying to other women, and thus likely to be associated with her long-term reproductive success. (It may also be significant that female orgasm is seen as gratifying to the man as well, and perhaps not just as a confirmation of his own sexual technique.)

It is worth noting that dominant male animals are typically less rushed and more deliberate about sexual intercourse, whereas social subordinates tend to be harried and thus hurried. Thus, we have watched dominant male grizzly bears copulate with sows in a manner that, if not altogether relaxed, at least indicates a degree of control, sexual no less than social. By contrast, subordinate male grizzlies spend much of their copulation time literally swiveling their heads over their shoulders, worrying about the imminent approach of dominant boars! There is no evidence that grizzly sows experience orgasm, but if they did, which type of boar would seem most likely to evoke such a response?

Sexual jealousy is a give-away. Its widespread existence suggests strongly that EPCs—that is, episodes of infidelity—have long been an important part of the human evolutionary past. There would be little reason for such a deep-seated tendency if it were not, to some extent, justified by events.

When a male bighorn sheep is alone with an estrous female, his courtship is likely to be comparatively slow and gentle; when rival males are present, the same male is likely to be more aggressive and brusque. Male rhesus monkeys will routinely attack females caught mating—or just con-sorting—with lower-ranking male rivals. Sometimes the females are severely injured. In one case, a female rhesus monkey who repeatedly approached another male was fatally injured by her top-ranking male consort. A male hamadryas baboon uses coercion to keep his small harem of females away from other males. If one of them happens to stray in the direction of other males, he threatens her with a conspicuous "eyebrow flash." If the errant female doesn't immediately mend her ways and approach the male, he will attack her with a vigorous neck-bite. There have been many other reports— especially for primates—of males herding their mates away from strange males, particularly during encounters with other groups.

A study of crab-eating macaques in captivity found that male aggression toward females occurred at the relatively low rate of once every three or four hours, so long as the individuals were housed in isolated pairs. When a rival male was introduced, the frequency increased to more than seven times per hour! There is a subtle difference of interpretation involved here. In the past, we assumed that male aggression toward females was simply intended to keep them away from horny and intrusive males. (The focus was on the behavior of *males*, whether seeking EPCs or concerned with mate-guarding.) Now, biologists have begun to see that such aggression is designed to prevent *females* from involving themselves with these other males. (The focus is increasingly on the behavior of *females* seeking EPCs.)

It is prominent among chimps as well: Jane Goodall reports that males are especially likely to "punish" a female who has been sexually involved with another male. Moreover, male chimpanzees sometimes use violence to force a female to follow them; they may spend considerable time herding a female away from other males, displaying substantial aggression toward her during that time. One male—Evered by name—spent five hours directing a female (Winkle) on a forced march, during which time he threatened her numerous times and physically attacked her five times, injuring her on two occasions. As the male increasingly gets his way, and the female is moved far from other males, he relaxes perceptibly. At the same time, the female— being more dependent on her persecutor/protector—typically becomes more cooperative and pliable. (The obvious human parallels are worrisome but, for all that, no less likely to be genuine.)

Regrettably—at least, by human standards—male chimps who are not sexually pushy and aggressive are generally less successful in consorting with females. In Goodall's account, an adult male named Jomeo was a perfect gentleman, showing the lowest rate of "punitive aggression" toward females. He was also the least successful when it came to forming consortships, and he appears to have been the only adult male who did not sire any offspring. Goodall speculates that males are often aggressive toward females in order to facilitate later sexual relationships: To the extent that a female is readily intimidated by given male, that male is more likely to obtain her sexual acquiescence in the future. Not a pretty picture, but one that comes more clearly into focus when we consider the role of

EPCs in predisposing males, in this case, to use aggression and even violence to force themselves upon females who might well have someone else in mind.

In short, it may be that some of the ugliest human behavior—marital abuse, wife beating, even homicide—derives at least in part from a widespread biological propensity to depart from monogamy.

Time now to summarize. (Easier said than done!) Human beings are unusual in their mating system. Although for the most part *Homo sapiens* is socially monogamous, displaying—for a mammal—huge amounts of paternal care, people also live almost colonially, often in enormous groups. In our social monogamy, we are like gibbons, yet we are also like chimps in that women interact regularly with other adults, not only other women but even other men. Sex, in such cases, may well be in the background, and for the most part that is where it stays. In some ways, we are more like certain colonial birds: socially monogamous yet rubbing shoulders with lots of other adults, every day. In such species, when males and females spend long periods of time apart from each other—one foraging, for example, while the other is attending the nest—both males and females have abundant opportunities for EPCs.

The human species is preferentially and biologically polygynous, but also mostly monogamous and—when conditions are ripe—avidly adulterous . . . all at once. There is no simple animal model that encompasses all of the "natural" human condition. Thus, in some species, males seek EPCs; in others, females do so. Which is the model for humans? Probably both.

Human beings use mate-guarding, frequent copulations, and also a hefty dose of social prescription—religious injunctions, cultural conditioning, legal restraints, eunuchs, chastity belts, female circumcision, and so forth—in efforts to impose their will (typically, the desires of powerful men) on everyone else's inclination. Rousseau speculated centuries ago that primitive human beings used to be happy, free, and socially independent of one another, but that most of our misfortune arose when the first people began identifying things—including sexual access to certain individuals—as their own. Maybe he was more right than most biologists have acknowledged, if many of the unpleasant aspects of male–male competition (notably, a penchant for violence) evolved because of the evolutionary payoff that comes with exclusive sexual access to one or more females. And with women inclined to accept, even seek, EPCs on occasion, conditions were ripe for the appearance of various increasingly competitive techniques whereby men tried to achieve sexual monopoly.

At the same time, it is not simply the churlishness of men or the voluptuous transgressions of women that would have opened the floodgates of original sin. If women really had no sexual urges beyond their designated mate, there would be very few EPCs; similarly, if other men were not willing, even eager, gallivanters. Neither men nor women are the primordial purveyors of EPC sinfulness, if sin it be. It takes two to do the EPC tango. And human beings love to dance.

# CHAPTER SEVEN

# So What?

A ccording to Saint Augustine, "The reason humans behave as they do is because they are not living in their true home." He meant God. Physical anthropologists mean that they are not living on the Pleistocene savanna! And students of animal mating systems could well mean that humans are not being permitted to live lives of polygyny or of monogamy plus EPCs.

But maybe our "true home" isn't such a nice place after all.

Western tradition makes it abundantly clear where it stands on monogamy and on adultery. The Sixth Commandment could hardly be more specific: "Thou shalt not commit adultery." And for good measure, the Tenth proclaims: "Thou shalt not covet thy neighbor's wife." The Old Testament is especially harsh on such transgressors; in Leviticus (20:10) and Deuteronomy (22:22) we learn that an adulteress and her lover are to be stoned. The New Testament, by contrast, is more forgiving, as evidenced by Jesus pardoning the woman taken in adultery, enjoining "Let he who is without sin cast the first stone." (Interesting to note: There are no comparably harsh penalties for a man, so long as he is not adulterous with another man's wife. Unmarried women, it seems, are fair game!)

Sigmund Freud once argued that the universality of the incest taboo suggests that incest avoidance is probably not instinctive, because, paradoxically, if it were, we would not need the restriction. We only need to be prohibited, the argument goes, from doing what we might otherwise attempt; there are no taboos against biting off one's own ears, for example. The persistent and explicit prohibitions against adultery in Western (and many other) traditions similarly confirm the biological arguments presented in this book; namely, that strict monogamy is not automatic. It needs to be enforced and reinforced. Otherwise, adultery happens.

Christianity is especially exercised about adultery. Jesus even inveighed against committing it in "one's heart," consistent with the fact that Christianity has historically taken a dim view of sex generally. In fact, sex is considered so degrading throughout much of the Christian tradition that marriage was widely acknowledged to be inferior to chastity. Marriage, in this view, exists only as a way of preventing the greater sin of fornication (defined as sex among unmarried people). As St. Paul put it, "It is good for a man not to touch a woman, nevertheless to avoid fornication let each man have his own wife and let each woman have her husband" (I Corinthians 7:1–2). According to Bertrand Russell, in *Marriage and Morals,*

> the Christian view that all intercourse outside marriage is immoral was . . . based upon the view that all sexual intercourse even within marriage is regrettable. A view of this sort which goes against biological facts can only be regarded by sane people as a morbid aberration.

According to John of Damascus, writing in the eighth century, Adam and Eve were created sexless; their sin in Eden led to the horrors of sexual reproduction. If only our earliest progenitors had obeyed God, we would be procreating less sinfully today (although it isn't at all clear how). "Matrimony is always a vice," claimed St. Jerome. "All that can be done is to excuse it and to sanctify it; therefore it was made a religious sacrament." Cleanliness is okay, but for many of the truly devout, celibacy was even closer to godliness. And sexual impurity was (and still is) dirty indeed. Christian monks up until the Renaissance complained bitterly of being visited in their sleep by succubi, female demons that gestured and beckoned lewdly to them, just as novitiate nuns were warned of the nocturnal visitations of their erotically enticing male counterparts, the incubi. For people who considered themselves married to Christ or to the Church, any sexual temptation—even if it involved nothing more than the occasional naughty dream or nocturnal ejaculation—was only slightly less sinful than outright fornication. The spirit may be willing, but the flesh can be adulterously weak.

And, of course, if marriage is fundamentally flawed, acceptable largely as a way of making sex tolerable (so long as it is within the marriage), then how much worse is marriage with explicitly prohibited (extramarital) sex thrown in?

Biblical tradition, however, was not nearly so uniformly sex-hating as the writings of early Christians might suggest. Old Testament men commonly had multiple wives, and some of the most reputable had numerous lovers as well as courtesans. The Song of Solomon is erotically charged—and so, apparently, was Solomon himself. Polygyny was widely accepted, and adultery was problematic only when it involved someone's wife or daughter; that is, a woman who was clearly associated with a man. Adultery was defined as a crime against a *man*, either husband or father . . . as it still is in much of the world today, especially regions influenced by Islam. Sexual relations between a married man and a woman who had neither a husband nor a father was not considered to violate any dictates, either of society or of God.

The Tenth Commandment does not say "Thou shalt not covet another woman." Rather, it is specifically concerned with protecting the rights of one's *neighbor*, by keeping gallivanting men away from other men's wives.

We may hear about loyal subjects or faithful servants—rarely about loyal rulers or faithful kings. "Ich dien" (I serve) is the motto of the Prince of Wales, but let's be honest: The prince is more likely to be served. Loyalty or fidelity is typically something that the strong demand of the weak. It may therefore surprise no one that the double standard demands fidelity from the wife, while typically winking at comparable actions by the husband. In ancient India, sex by a married man with a prostitute or slave woman wasn't adultery, unless she was someone else's property, in which case it was an offense against the owner, not against the woman herself, and certainly not against the Lothario's wife. It is worth noting, incidentally, that male-oriented sexual ethics of this sort do not necessarily imply a rigid prudishness on the part of society at large: India, for example, is home of the world's first and most detailed sex manual, the *Kama Sutra,* and Indian lore has long glorified the pleasures of sex, to the point that Shiva and his wife were sometimes pictured as prolonging intercourse almost to infinity.

Among the ancient Hebrews, at least, there were other reasons for enforcing a double standard. Marriage was especially important as a means of establishing the property rights of geneological succession. Thus, an adulterous wife disrupted the careful system of biological relatedness upon which the social network depended.

Generally, Protestantism has been more agitated about adultery than has Catholicism, probably because the latter prohibits divorce and remarriage—at least without an annulment. Thus, when divorce is very difficult to obtain because of religious reasons, extramarital affairs are less threatening to the continuation of the marriage. A wife may feel insulted, belittled, and generally furious at her philandering husband, but at least she is less likely to find her marriage terminated as a result. Her status as wife remains relatively unchallenged. (On the other hand, she might want to be relieved of such a spouse, but reluctant to go through the time and expense involved in an annulment.) By contrast, the availability of divorce among Protestants has upped the ante, making it possible that an extramarital affair can have more serious consequences. The American Puritans were especially severe in their punishment of adultery, which for a time was a capital crime in both the Massachusetts and Plymouth colonies.

It is noteworthy that the usually puritanical Puritans wrote enthusiastically about the erotic joys of marriage, albeit in coded language. For example, John Milton, in *Paradise Lost:*

Hail wedded Love, mysterious Law, true source
Of human offspring, sole propertie,
In Paradise of all things common else.
By thee adulterous lust was driv'n from men
Among the bestial herds to range . . .

In the Western world, sex and marriage have traditionally been closely bound—so closely, in fact, that the former was defined by the latter. So we have premarital sex, marital sex, and extramarital sex. (There are no words yet, interestingly, for post-divorce sex or even widow or widower sex. Would a sexually active confirmed bachelor, age 45, be engaging in "premarital" sex?) Sex itself may be in the process of defining itself independent of marriage. Whether this is healthy, however, is another question. Presumed experts have come down hard on both sides of the issue:

Thus, psychologist Havelock Ellis has written:

The man who resides in a large urban area and who never once, during thirty or more years of married life, is sorely tempted to engage in adultery for purposes of sexual variety is to be suspected of being indeed biologically and/or psychologically abnormal; and he who frequently has such desires and who occasionally and unobtrusively carries them into practice is well within the normal healthy range.

Alternatively, many marriage counselors and psychotherapists, as well as classical psychoanalysts, view extramarital affairs as neurotic at best, resulting from narcissism, character disorders, fragmentary superegos, infantile love-needs, and the like. In their book *The Wandering Husband,* H. Spotnitz and L. Freeman maintain that "Infidelity may be statistically normal but it is also psychologically unhealthy. . . . It is a sign of emotional health to be faithful to your husband or wife."

The human being is a complex creature. He and she live within an elaborate framework of cultural prescriptions, biological inclinations, historical traditions, psychological processes, and personal experiences. So what if monogamy is not natural? And if adultery is? Can anything human be natural? Can anything human be unnatural? And what difference does it make?

"All tragedies are finish'd by a death," wrote Byron, whereas "All comedies are ended by a marriage." But whether or not monogamous marriage is our "natural" state, the human comedy rarely ends with it; more often, marriage is only a beginning.

"Forsaking all others," say the wedding vows of almost every Jewish or Christian denomination. Fail in this promised forsaking and it can be very costly, not only in money but in career, peace of mind, one's marriage and family, self-esteem, and the esteem of others. Sometimes extramarital affairs are hushed up, even in high places, as were John F. Kennedy's many infidelities in the White House. At other times, they become public and are ruinous, as with the great nineteenth-century Irish leader Parnell, who nearly brought independence to his country, only to be publicly shamed and politically discredited when his affair with a married woman, Mrs. Kitty O'Shea, was revealed.

And sometimes extramarital affairs become front-page news, complete with lurid details, denials, recantings, impeachment, and then, despite the shame, a degree of exoneration.

In *Too Far to Go,* John Updike wrote that marriage is a "million mundane moments shared." Without love, that sharing would presumably not be sought in the first place; and out of that sharing, love matures and grows. However, the sharing of a million mundane moments can also get pretty boring. Holy wedlock can become holy deadlock. As an anonymous Greek author wrote more than 2,000 years ago:

Once plighted, no men would go awhoring;
They'd stay with the ones they adore,
If women were half as alluring
After the act as before.

According to Denis de Rougemont, there is an "inescapable conflict in the West between passion and marriage." Our civilization must recognize, he urges, "that marriage, upon which its social structure stands, is more serious than the love which it cultivates, and that marriage cannot be founded on a fine ardor."

The issue, for historian de Rougemont, is the *danger* of passion: We adore passion, and we are fascinated by it. According to de Rougemont, we even have a perverse desire to achieve unhappiness, to attain tragic proportions:

Western Man is drawn to what destroys "the happiness of the married couple" at least as much as to anything that ensures it. Where does this contradiction come from? If the breakdown of marriage has been simply due to the attractiveness of the forbidden, it still remains to be seen why we hanker after unhappiness, and what notion of love—what secret of our existence, of the human mind, perhaps of our history—this hankering must hint at.

Maybe some people go outside monogamy precisely so that they will be caught and punished—and thus achieve access to the romantic, the intense, and the tragic: "whether in fact or in dreams, in remorse or in terror, in the delight of revolt or the disquiet of temptation." In any event, many would ruefully agree with Alexandre Dumas (the younger) that "the chains of marriage are so heavy that it takes two to bear them, and sometimes three."

All the world loves a lover, and the more deeply he or she loves, the better. Yet we don't normally speak of a passionate marriage. A good marriage, a happy marriage, a comfortable and compatible marriage, yes, but only rarely a passionate one. Or at least, not for long. "They lived happily ever after." Sure. But "They lived passionately ever after"? Come on.

If nothing else, it would be exhausting. To live in a state of perpetual passion would be to forgo much of the rest of life, and, in truth, there *are* other things. Love can deepen and broaden, provide new areas of connectedness and new strengths, but it rarely becomes more passionate. Even the simplest animals, the protozoa, are subject to the most primitive form of learning, habituation. An animal habituates to something when it stops responding to it or responds less than it used to. We habituate to smells, to sounds, even to sights, as when we stop seeing the paintings or photos on our wall. One way to counter habituation is to change the stimulus: You may have habituated to the sound of the refrigerator motor, but when it turns itself off, or changes pitch, you suddenly notice it again. Some people periodically rearrange the art on their walls so as to appreciate it afresh. Rearranging our love lives, however, is a different matter.

Passion is by definition short lived, or at most medium lived. Almost never is it long-lasting. It flourishes when new and freshly kindled, or perhaps when it is forbidden, as in the case of Romeo and Juliet and, of course, adultery. It also gets a shot in the arm from a change, when directed toward someone new (or, as any experienced—and happily—married couple knows, when the familiar person is experienced in a different way).

De Rougemont maintains that there are two moralities: one of marriage, and the other of passion, and that every married person must choose. The adulterer seeks to have both, but with different partners. In the great French novel of infidelity, *Madame Bovary,* Flaubert's heroine became unfaithful when life with her dull doctor husband became tedious and she found herself wondering if there wasn't more to life than "this."

If marriage is in a sense the cradle of adultery, is it therefore also the grave of love? Not at all, or, at least, not necessarily. As philosopher Benedetto Croce puts it (and de Rougemont would doubtless agree), marriage is rather "the grave of *savage* love." Others, seeking a constantly rekindled savage love, periodically go outside their marriage looking for the tinder of renewed savagery.

As Freud pointed out in *An Outline of Psychoanalysis,* erotic dreams rarely involve one's spouse; conscious extramarital sexual imaginings are pretty much universal. (Recall that even straight-laced presidential candidate Jimmy Carter admitted in a controversial interview with *Playboy* magazine that he had occasionally committed "lust" in his heart.) But perhaps it is precisely when—and because—the flesh is weak that the spirit ought to rise to the occasion. Thus, in *Beyond the Pleasure Principle,* Freud also suggested that we all experience an ongoing struggle between the "pleasure principle," which includes sexual activity in particular and which constantly seeks gratification, and the "reality principle," expressed by the superego or, in simple terms, the conscience.

Moreover, although "what comes naturally" is—almost by definition—easy to do, this doesn't mean that it is right. The crowning glory of *Homo sapiens* is its huge brain. This remarkable organ gives people the ability, perhaps unique in the living world, to reflect on their inclinations and decide, if they choose, to act contrary to them. In Mozart's opera *The Marriage of Figaro,* we are advised, "Drink when you are not thirsty, make love when you don't want to this is what distinguishes us from the beasts." What about *not* making love when we *do* want to? There may be no way to affirm one's humanity as effectively as by saying "no" to some deeply held predispositions, especially when that wonderful brain of ours advises that such predispositions may be troublesome, for ourselves or others.

Not many people reflect on a spouse's adultery with the comical, cerebral detachment of Leopold Bloom in James Joyce's *Ulysses:*

> As natural as any and every natural act of a nature expressed or
> understood executed in natured nature by natural creatures in
> accordance with his, her and their natured natures, of dissimilar
> similarity. As not as calamitous as a cataclysmic annihilation of the

planet in consequence of collision with a dark sun. As less reprehensible than theft, highway robbery, cruelty to children and animals, obtaining money under false pretenses, forgery, embezzlement, misappropriation of public money, betrayal of public trust, malingering, mayhem, corruption of minors, criminal libel, blackmail, contempt of court, arson, treason, felony, mutiny of the high seas, trespass, burglary, jailbreaking, practice of unnatural vice, desertion from armed forces in the field, perjury, poaching, usury, intelligence with the king's enemies, impersonation, criminal assault, manslaughter, willful and premeditated murder. As not more abnormal than all other altered processes of adaptation to altered conditions of existence, resulting in a reciprocal equilibrium between the bodily organism and its attendant circumstances, foods, beverages, acquired habits, indulged inclinations, and significant disease.

But then Mr. Bloom, whose wife had an afternoon rendezvous with her hot new lover, the aptly named Blazes Boylan, concludes his reverie by noting that Molly's affair is "more than inevitable, irreparable." We have already seen that human beings are not, biologically speaking, monogamous. But they are also disinclined to tolerate departures from monogamy with the blithe assurance that because they are "natural," they are okay. At the same time, Leopold Bloom is wrong: Marital infidelity is not inevitable (nor, one hopes, necessarily irreparable either).

For human beings, sex has three great functions: procreational, relational, and recreational. The first is obvious. The second speaks to the deep bonding and connectedness that often develop between lovers and that—according, at least, to Western religious tradition—should precede sexual relations between people. The third aspect of sex, recreational, is doubtless the most controversial. But the fact remains that sex is, or can be, great fun and a powerful recreational urge in its own right.

In *Ars Amatoria*, the Roman poet Ovid justifies what is perhaps the most notorious, and ruinous, of all cases of adultery: Helen's affair with Paris, which precipitated the Trojan War and "launched a thousand ships." It seems that Helen's husband Menelaus was away at the time:

Afraid of lonely nights, her spouse away
Safe in her guest's warm bosom, Helen lay.
What folly, Menelaus, forth to wend,
Beneath one rooftree leaving wife and friend? . . .
Blameless is Helen, and her lover too:
They did what you or anyone would do.

Adultery is hot stuff, emotionally charged to a degree that must seem remarkable to anyone not bringing a biological perspective to the human condition. During the famous Kinsey sex studies, for example, the single largest cause of people's refusing to participate was a question about extramarital sex.

In *Civilization and Its Discontents,* Freud suggested that civilization is built on the repression of the instincts. And we now know that one of these instincts apparently leans toward multiple matings. Civilization is presumably facilitated by controlling antisocial tendencies such as murder, rape, or robbery. Is there anything antisocial about multiple matings? Yes—if society forbids such behavior and if monogamy is contracted for and, hence, expected by the other spouse. There is much to be said for plain old-fashioned honesty and integrity; a creature so cerebral as to be able to establish elaborate rules and expectations for domestic life (not to mention the pursuit of science, literature, art, and so forth) should also be capable of keeping his or her word.

Civilization is founded not only on the repression of the instincts but also on the ascendancy of law. (In some ways, the two are synonymous.) Therefore, once a law or social expectation exists, unless it is grossly unfair there is a presumption that decency and social order are furthered by obedience to it. Once a monogamous code exists, therefore, perhaps violating that code is antithetical to higher levels of civilization and of personal development. Notably, however, anthropologists have found no correlation between extramarital restrictiveness and a society's level of social complexity. (Premarital restrictiveness, on the other hand, tends to be greater in more complex, although not necessarily "better," societies.)

Many "advanced" civilizations were polygynous, and some simple, nontechnological ones, strictly monogamous. Even if monogamy isn't necessary for civilization, however, it is clear that public adherence to monogamous ideals is necessary for success and survival in *current* Western civilization, if only because that is the way we have defined ourselves. It is hard for a public bigamist or adulterer to "get ahead." Ask former presidential candidate Gary Hart. And even those at the pinnacle of power can be toppled or severely tarnished. Ask Bill Clinton, or even Newt Gingrich.

As to the procrustean nature of monogamy inhibiting some of our deeper inclinations, isn't that what growing up is all about? After all, as we get older, we are all expected to do what is permitted and inhibit what is forbidden: we learn toilet training, not to hit or bite, to say *please* and *thank you,* and generally to rein ourselves in. Many things are natural but unpleasant: trichinosis, warts, hurricanes. So even if enforced monogamy is in some sense "unnatural," this doesn't necessarily mean that it is undesirable. (It also doesn't have to be unpleasant . . . but that is another story, and perhaps another book!) Those animal species that are reliably monogamous— a declining list—are sexually faithful because they have no real choice. But people do. In this regard, our biology is neither overweening nor even a worthwhile guide. Certainly it is no excuse.

Whatever our natural inclinations, there is no doubt that human beings are biologically and psychologically capable of having sex with more than one person, often in fairly rapid succession. The evidence is also overwhelming that many people are capable not only of "making love to" but also of loving more than one person at the same time. But we are socially prohibited from doing either. This social prohibition is a powerful one, and in the long run, it generally wins, although usually not without a struggle and often with some short-term defeats. And that struggle—experienced as occasional brief flings for a night or a weekend, long extramarital relationships over months or years, or just fantasized encounters—may be the source of some of the most complex, intense, and confusing emotions that human beings experience.

It has been suggested that the mental health profession often serves as a social Band-Aid, simply helping people adjust to a sick society and often siphoning energy and attention from where it is most needed: the reformation of social ills. Perhaps the effort expended in adjusting to monogamy is like this. Perhaps we should instead adjust our ideals of monogamous matrimony to accord with human inclinations. Maybe instead of taking monogamy as the norm, and thus being "shocked, shocked" by adultery— like the notorious police captain in the movie *Casablanca*—we should see infidelity as the baseline condition, whereupon we might be free to examine monogamy, dispassionately, for the rarity that it is.

This assumes, however, that there is a better alternative, for example, that open, unstructured, and nonrestrictive sexual relationships would make people happier. There is no reason to believe that this is true. Indeed, many "utopian" social experiments have failed precisely because feelings of interpersonal possessiveness got in the way of the idealized dream of social and sexual sharing.

Society, with its expectation of monogamy, establishes the boundaries of who is and who is not an acceptable sexual partner. Marriage, it is assumed, narrows this field considerably: to just one individual. Some may find this stultifying, others reassuring, since it generates a haven of surety and confidence, a womb with a view, ideally free of the brawls of sexual competition.

No other marital pattern—polygyny, polyandry, group marriage, "open" marriage—has been shown to work better. Nonetheless, monogamy does not work perfectly, and throughout history people have been delighted and bedeviled, energized and agonized, by monogamy and by departures from it. On balance, perhaps monogamy is like Winston Churchill's description of democracy: the worst possible system, except when you consider the alternatives.

Maybe living things—or, at least, the lucky ones—are somehow destined to achieve perfect one-to-one relationships; that is, perhaps everyone has a true soul-mate out there somewhere. The only question is whether these two halves of a potentially perfect whole will succeed in finding each other, á la Plato's tongue-in-cheek version. Don't bet on it.

This is not to say that monogamy—even happy, fulfilled monogamy—is impossible, because, in fact, it is altogether within the realm of human possibility. But since it is not natural, it is not easy. Similarly, this is not to say that monogamy isn't desirable, because there is very little connection, if any, between what is natural or easy and what is good.

But let us imagine, just for argument's sake, that Plato was factually correct: that for each of us, there exists the perfect mate, the ideal counterpart, the hand-in-glove Siamese twin with whom we would be perfectly in love and eternally happy. There are 6 billion people on our planet, of whom we meet probably fewer than several thousand in a lifetime. This works out to about one in a million. Accordingly, for every person we meet, there are about 999,999 we never do. And of those few we actually do meet, only a small proportion of those encounters occur for us at ages and in circumstances in which love and/or marriage—never mind sex—are even feasible. In short, the chances are pretty slim that we will ever meet our perfect other half, even if he or she exists.

But don't despair! The future is not necessarily bleak, neither for personal happiness nor even for monogamy itself (assuming, of course, that one is sufficiently committed, at least to the latter). Even though there may be no perfect other half ideally crafted for each person—just waiting to be thrown together by fate, some enchanted evening—in the course of a loving marriage, two people have the opportunity to hone and shape their shared experiences such that one's partner does in fact become a rather precisely fitting key, uniquely adapted to the other's lock, and vice versa. The perfect fit of a good monogamous marriage is made, not born. And despite the fact that much of our biology seems to tug in the opposite direction, such marriages can in fact be made. It is an everyday miracle.

Monogamy among animals is a matter of biology. So is monogamy among human beings. But in the human case, monogamy is more. It is also a matter of psychology, sociology, anthropology, economics, law, ethics, theology, literature, history, philosophy, and most of the remaining humanities and social sciences as well. Not to mention love, trust, hope, disillusionment, fear, anger, frustration, disappointment, delight, despair, amusement, irritation, expectation, tolerance, intolerance, loyalty, betrayal, lust, boredom, excitement, confidence, anxiety, money, poverty, children, barrenness, engagement, disengagement, sickness, health, life, and death. And just about everything else.

# Notes

## CHAPTER 1 Monogamy for Beginners

**2** Marriage is the ultimate sanction: Or, at least, the ultimate interpersonal sin liable to be experienced by many people; adulterers outnumber murderers by a hefty margin.

**3** On the other hand, there are cases: J.-G. Baer and L. Euzet. 1961. Classe de Monogenes. In *Traite de Zoologie, Tome* IV, ed. P.-P. Grasse. Paris: Masson et Cie.

**5** In 1970, in what can truly be called: G.A. Parker. 1970. Sperm competition and its evolutionary consequences in the insect. *Biological Reviews* 45: 525–567.

**6** Here is an account of sperm competition [subsequent quote]: T. R. Birkhead and G. A. Parker. 1997. Sperm competition and mating systems. In *Behavioural Ecology: An Evolutionary Approach,* ed. J. R. Krebs and N. B. Davies. Oxford: Blackwell Science.

**7** It required the next and most significant breakthrough: A. J. Jeffreys, V. Wilson, and S. L. Thein. 1985. Hypervariable "minisatellite" regions in human DNA. *Nature* 314: 67–73.

**8** It is presented not to provide [subsequent quote]: J. G. Ewen, D. P. Armstrong, and D. M. Lambert. 1999. Floater males gain reproductive success through extrapair fertilizations in the stitchbird. *Animal Behaviour* 58: 321–328.

**10** There is also strong evidence: M. Morris. 1993. Telling tales explains the discrepancy in sexual partner reports. *Nature* 365: 437–440.

**10** Increasingly, biology journals: To avoid any misunderstanding, note that "tits" are to British ornithologists what chickadees are to biologists on this side of the Atlantic.

**10** We have even had this oxymoronic report: D. E. Gladstone. 1979. *The American Naturalist* 114: 545–547.

**11** And this *before* having reached: W. B. Quay. 1985. Cloacal sperm in spring migrants: occurrence and interpretation. *Condor* 87: 273–280.

**11** Similarly, there is no guaranteed correlation: H. L. Gibbs, P. J Weatherhead, P. T. Boag, B. N. White, L. M. Tabak, and D. J. Hoysak. 1990. Realized reproductive success of polygynous red-winged blackbirds revealed by DNA markers. *Science* 250: 1394–1397.

**12** Even females of seemingly solitary species: P. S. Rodman and J. C. Mitani. 1987. Orangutans: sexual dimorphism in a solitary species. In *Primate Societies*, ed. B. B. Smuts, D. L. Cheney, R. M. Seyfarth, R. W. Wrangham, and T. T. Struhsaker. Chicago: University of Chicago Press; A. Schenk and K. M. Kovacs. 1995. Multiple mating between black bears revealed by DNA fingerprinting. *Animal Behaviour* 50: 1483–1490.

**12** More than 65 percent: R. A. Mulder, P. O. Dunn, A. Cockburn, K. A. Lazenby-Cohen, and M. J. Howell. 1994. Helpers liberate female fairy-wrens from constraints on extra-pair mate choice. *Proceedings of the Royal Society of London, Series B* 225: 223–229.

**12** Warblers and tree swallows: P. O. Dunn and R. J. Robertson. 1993. Extra-pair paternity in polygynous tree swallows. *Animal Behaviour* 45: 231–239; K. Schulze-Hagen, I. Swatschek, A. Dyrcz, and M. Wink. 1993. Multiple Vaterschaften in Bruten des Seggenrohrsangers *Acrocephalus paludicola*: erste Ergebnisse des DNA-Fingerprintings. *Journal of Ornithology* 134: 145–154.

**12** Even prior to DNA fingerprinting: J. H. Edwards. 1957. A critical examination of the reputed primary influence of ABO phenotype on fertility and sex ratio. *British Journal of Preventive and Social Medicine* 11: 79–89.

**12** In response to surveys: E. O. Laumann, J. H. Gagnon, R. T. Michael, and S. Michaels. 1994. *The Social Organization of Sexuality*. Chicago: University of Chicago Press.

**13** The numbers for women are a bit lower: A. C. Kinsey, W. B. Pomeroy, C. E. Martin, and P. H. Gebhard. 1953. *Sexual Behavior in the Human Female*. Philadelphia: W. B. Saunders.

**14** So polyandry, which means "many males": Etymological note: The word *polygamy* is often used incorrectly when *polygyny* is more appropriate. *Polygamy* literally means "many gametes" and is more precisely either polygyny or polyandry. People who speak of Mormon polygamy, for example, or polygamy as described in the Bible or as practiced in current Islamic and some African societies really mean polygyny, or harem-keeping.

## CHAPTER 2 Undermining the Myth: Males

**17** At the same time, females: G. C. Williams. 1966. *Natural Selection and Adaptation*. Princeton, NJ: Princeton University Press.

**17** An important conceptual breakthrough: R. L. Trivers. 1972. Parental investment and sexual selection. In *Sexual Selection and the Descent of Man, 1871–1971*, ed. B. Campbell. Chicago: Aldine; see also T. H. Clutton-Brock and A. C. J. Vincent. 1991. Sexual selection and the potential reproductive rates of males and females. *Nature* 351: 58–60.

**18** Under the pressure of sperm competition: G. A. Parker. 1982. Why are there so many tiny sperm? Sperm competition and the maintenance of two sexes. *Journal of Theoretical Biology* 96: 281–294.

**20** The function of these giant *Drosophila* sperm: S. Pitnick, G. S. Spicer, and T. A. Markow. 1995. How long is a giant sperm? *Nature* 375: 109.

**21** As to human beings [subsequent quote]: I. Schapera. 1940. *Married Life in an African Tribe*. London: Faber & Faber.

**21** Indeed the famous team of sex researchers [subsequent quote]: A. C. Kinsey, W. B. Pomeroy, and C. E. Martin. 1948. *Sexual Behavior in the Human Male.* Philadelphia: W. B. Saunders.

**22** In addition, although the Coolidge effect: D. A. Dewsbury. 1981. An exercise in the prediction of monogamy in the field from laboratory data on 42 species of muroid rodents. *Biologist* 63: 138–162; A. F. Dixson. 1995. Sexual selection and ejaculatory frequencies in primates. *Folia Primatologica* 64: 146–152.

**23** Also, males having a high reproductive success: H. L. Gibbs, P. J. Weatherhead, P. T. Boag, B. N. White, L. M. Tabak, and D. J. Hoysak. 1990. Realized reproductive success of polygynous red-winged blackbirds revealed by DNA markers. *Science* 250: 1394–1397.

**24** In fact, two of the unpaired floaters: John G. Ewen, D. P. Armstrong, and D. M. Lambert. 1999. Floater males gain reproductive success through extrapair fertilizations in the stitchbird. *Animal Behaviour* 58: 321–328.

**24** Regardless of the mechanism: T. R. Birkhead, J. E. Pellatt, and F. M. Hunter. 1988. Extra-pair copulation and sperm competition in the zebra finch. *Nature* 334: 60–62.

**25** But, instead, EPCs: A. P. Møller. 1998. Sperm competition and sexual selection. In *Sperm Competition and Sexual Selection,* ed. T. R. Birkhead and A. P. Møller. San Diego: Academic Press.

**25** Now, DNA fingerprinting shows: J. M. Pemberton, S. D. Albon, F. E. Guinness, T. H. Clutton-Brock, and G. A. Dover. 1992. Behavioral estimates of male mating success tested by DNA fingerprinting in a polygynous mammal. *Behavioral Ecology* 3: 66–75.

**25** In one study involving birds: T. R. Birkhead, F. Fletcher, E. J. Pellatt, and A. Staples. 1995. Ejaculate quality and the success of extra-pair copulations in the zebra finch. *Nature* 377: 422–423.

**26** So, one possible explanation: M. Kirkpatrick, T. Price, and S. J. Arnold. 1990. The Darwin-Fisher theory of sexual selection in monogamous birds. *Evolution* 44: 180–193.

**26** It provides strong evidence: B. C. Sheldon and J. Ellegren. 1999. Sexual selection resulting from extrapair paternity in collared flycatchers. *Animal Behaviour* 57: 285–298.

**27** There would then be a reproductive payoff: H. Ellegren, L. Gustafsson, and B. C. Sheldon. 1996. Sex ratio adjustment in relation to paternal attractiveness in a wild bird population. *Proceedings of the National Academy of Sciences* 93: 723–728.

**27** A typical one: B. Kempenaers, G. R. Verheyen, M. Van den Broeck, T. Burke, C. Van Broeckhoven, and A. A. Dhondt. 1992. Extra-pair paternity results from female preference for high-quality males in the blue tit. *Nature* 357: 494–496; S. M. Yezerinac, P. J. Weatherhead, and P. T. Boag. 1995. Extra-pair paternity and the opportunity for sexual selection in a socially monogamous bird (*Dendroica petechia*). *Behavioral Ecology and Sociobiology* 37: 179–188.

28 It appears that older, more colorful males: H. G. Smith, R. Montgomerie, T. Poldmaa, B. N. White, and P. T. Boag. 1991. DNA fingerprinting reveals relation between tail ornaments and cuckoldry in barn swallows. *Behavioral Ecology* 2: 90–98; D. Hasselquist, S. Bensch, and T. von Schantz. 1996. Correlation between male song repertoire, extra-pair paternity and offspring survival in the great reed warbler. *Nature* 381: 229–232; J. Sundberg and A. Dixon. 1996. Old, colourful male yellowhammers, *Emberiza citrinella*, benefit from extra-pair copulations. *Animal Behaviour* 52: 113–122.

28 Among cattle egrets: M. Fujioka and S. Yamagishi. 1981. Extramarital and pair copulations in the cattle egret. *Auk* 9: 134–144.

28 Similarly, male secondary sexual traits: G. E. Hill, R. Montgomerie, T. Roeder, and P. T. Boag. 1994. Sexual selection and cuckoldry in a monogamous songbird: implications for sexual selection theory. *Behavioral Ecology and Sociobiology* 35: 193–199; O. Ratti, M. Hovi, A. Lundberg, H. Tegelstrom, and R. V. Alatalo. 1995. Extra-pair paternity and male characteristics in the pied flycatcher. *Behavioral Ecology and Sociobiology* 37: 419–425; P. J. Weatherhead and P. T. Boag. 1995. Pair and extra-pair mating success relative to mate quality in red-winged blackbirds. *Behavioral Ecology and Sociobiology* 37: 81–91.

29 But occasionally the in-pair female: T. H Birkhead, T. Burke, R. Zann, F. M. Hunter, and A. P. Krupa. 1990. Extra-pair paternity and intraspecific brood parasitism in wild zebra finches, *Taeniopygia guttata*, revealed by DNA fingerprinting. *Behavioral Ecology and Sociobiology* 27: 315–324.

29 Guarding takes about a quarter of an hour: G. A. Parker. 1970. The reproductive behaviour and the nature of sexual selection in *Scatophaga stercoraria* L. (Diptera: Scatophagidae). IV. The origin and evolution of the passive phase. *Evolution* 24: 774–788.

29 If another male dungfly: H. Sigurjonsdottir and G. A. Parker. 1981. Dung fly struggles: evidence for assessment strategy. *Behavioral Ecology and Sociobiology* 8: 219–230.

30 Male bank swallows, for example: M. D. Beecher and I. M. Beecher. 1979. Sociobiology of bank swallows: reproductive strategy of the male. *Science* 205: 1282–1285.

30 Mate-guarding is also a common male strategy: P. W. Sherman. 1989. Mate guarding as paternity insurance in Idaho ground squirrels. *Nature* 338: 418–420.

30 A now-classic anthropological review: G. P. Murdock. 1967. *Culture and Society*. Pittsburgh, PA: University of Pittsburgh Press.

30 In some societies, husbands: R. Benedict. 1934. *Patterns of Culture*. Boston: Houghton Mifflin.

30 Such concern may not be ill founded: R. Baker and M. Bellis. 1995. *Human Sperm Competition*. London: Chapman & Hall.

30 Here is a description of courtship [subsequent quote]: E. Selous. 1933. *Evolution of Habit on Birds*. London: Constable.

31 If there is a continuing arms race: For some of the range of correlations—neutral, positive and negative—see S. B. Meek, R. J. Robertson, and P. T. Boag. 1994. Extrapair paternity and intraspecific brood parasitism in eastern bluebirds as revealed by DNA fingerprinting. *Auk* 111: 739–744; B. Kampenaers, G. R. Verheyen, and A. A. Dhondt. 1995. Mate guarding and copulation behaviour in monogamous and polygynous blue tits: do males follow a best-of-a-bad-job strategy? *Behavioral Ecology and Sociobiology* 36: 33–42; T. Burke et al. 1989. Parental care and mating behaviour of polyandrous dunnocks *Prunella modularis* related to paternity by DNA fingerprinting. *Nature* 338: 249–251.

31 Among a group of insects: G. A Gangrade. 1963. A contribution to the biology of *Necroscia sparaxes* Westwood (Phasmidae: Phasmida). *Entomologist* 96: 83–93.

31 Thus, it is typically more intense: E.g., J. L Dickinson and M. L. Leonard. 1997. Mate-attendance and copulatory behaviour in western bluebirds: evidence of mate guarding. *Animal Behaviour* 52: 981–992.

31 Among barn swallows: A. P. Møller. 1987. Extent and duration of mate guarding in swallows *Hirundo rustica*. *Ornis Scandinavica* 18: 95–100.

32 Females of this species breed in alternate years: D. P. Barash. 1981. Mate-guarding and gallivanting by male hoary marmots (*Marmota caligata*). *Behavioral Ecology and Sociobiology* 9: 187–193.

32 As soon as their mates are infertile: L. M. Brodsky. 1988. Mating tactics of male rock ptarmigan *Lagopus mutus*: a conditional strategy. *Animal Behaviour* 36: 335–342.

32 Just about always: e.g., J. L. Dickinson. 1997. Male detention affects extra-pair copulation frequency and pair behavior in western bluebirds. *Animal Behaviour* 51: 27–47; D. F. Westneat. 1994. To guard or go forage: conflicting demands affect the paternity of male red-winged blackbirds. *The American Naturalist* 144: 343–354.

32 By contrast, when males were removed: D. Currie, A. P. Krupa, T. Burke, and D. B. A. Thompson. 1999. The effect of experimental male removals on extrapair paternity in the wheatear, *Oenanthe oenanthe*. *Animal Behaviour* 57: 145–152.

34 The possibility arises that adult males: I. U. Reichard. 1995. Extra-pair copulations in a monogamous gibbon (*Hylobates lar*). *Ethology* 100: 99–112.

34 As a result, the younger males fathered: E. S. Morton, L. Forman, and M. Braun. 1990. Extrapair fertilizations and the evolution of colonial breeding in purple martins. *Auk* 107: 275–283.

35 Males that are cuckolded: P. C. Frederick. 1987. Extrapair couplations in the mating system of white ibis (*Eudocimus albus*). *Behaviour* 100: 170–201; D. F. Westneat. 1988. Parental care and extrapair copulations in the indigo bunting. *Auk* 105: 149–160.

35 The general pattern is concisely described: A. Johnsen and J. T. Lifjeld. 1995. Unattractive males guard their mates more closely: an experiment with bluethroats (Aves, Turdidae: *Luscinia s. svecica*). *Ethology* 101: 200–212.

35 Several studies have confirmed: B. Kempenaers, G. R. Verheyen, M. Van den Broeck, T. Burke, C. Van Broeckhoven, and A. A. Dhondt. 1992. Extra-pair paternity results from female preference for high-quality males in the blue tit. *Nature* 357: 494–496; B. Kempenaers, G. R. Verheyen, and A. A. Dhondt. 1995. Mate guarding and copulation behaviour in monogamous and polygynous blue tits: do males follow a best-of-a-bad-job strategy? *Behavioral Ecology and Sociobiology* 36: 33–42.

35 And not only birds: S. W. Gangestad and R. Thornhill. 1997. An evolutionary psychological analysis of human sexual selection: developmental features, male sexual behavior, and mediating features. In *Evolutionary Social Psychology*, ed. J. A. Simpson and D. T. Kenrick. Hillsdale, NJ: Erlbaum.

35 Several times this was seen: A. P. Møller. 1990. Deceptive use of alarm calls by male swallows *Hirundo rustica*: a new paternity guard. *Behavioral Ecology* 1: 1–6.

36 By ruining her web: P. J. Watson. 1986. Transmission of a female sex pheromone thwarted by males in the spider *Linyphia litigiosa* Keyserling (Linyphiidae). *Science* 233: 219–221.

36 Eberhard reviewed the sweaty details: W. G. Eberhard. 1994. Evidence for widespread courtship during copulation in 131 species of insects and spiders, and implications for cryptic female choice. *Evolution* 48: 711–733.

37 Because of their basic biology: D. Dewsbury. 1982. Ejaculate cost and male choice. *The American Naturalist* 119: 601–610.

37 Sure enough, when male rats: R. R. Bellis, M. A. Baker, and M. J. G. Gage. 1990. Variation in rat ejaculates consistent with the kamikaze sperm hypothesis. *Journal of Mammalology* 71: 479–480.

37 Plains zebra stallions copulate less: J. R. Ginsberg and D. I. Rubenstein. 1990. Sperm competition and variation in zebra mating behavior. *Behavioral Ecology and Sociobiology* 26: 427–434.

38 Males are absent: T. Birkhead and C. M. Lessels. 1988. Copulation behaviour of the osprey *Pandion haliaetus*. *Animal Behaviour* 36: 1672–1682.

38 This has been especially well established: A. Poole. 1989. *Ospreys: A Natural and Unnatural History*. Cambridge: Cambridge University Press.

38 Among orioles, males will copulate: B. B. Edinger. 1988. Extra-pair courtship and copulation attempts in northern orioles. *Condor* 90: 546–554.

**39** Very quickly afterward: D. P. Barash. 1977. Sociobiology of rape in mallards (*Anas platyrhinchos*): responses of the mated male. *Science* 197: 788–789.

**39** As soon as one male copulates: J. Faaborg and J. C. Bednarz. 1990. Galápagos and Harris' hawks: divergent causes of sociality in two raptors. In *Cooperative Breeding in Birds*, ed. P. B. Stacey and W. D. Koenig. Cambridge: Cambridge University Press.

**39** Among rats, males mate: M. K. Matthews and N. T. Adler. 1977. Systematic interrelationship of mating, vaginal plug position, and sperm transport in the rat. *Physiology and Behavior* 20: 303–309.

**39** Among nonhuman primates: C. D. Busse and D. Q. Estep. 1984. Sexual arousal in male pigtailed monkeys (*Macaca nemestrina*): effects of serial matings by two males. *Journal of Comparative Psychology* 98: 227–231; D. Q. Estep, T. P. Gordan, M. E. Wilson, and M. L. Walker. 1986. Social stimulation and the resumption of copulation in rhesus (*Macaca mulatta*) and stumptail (*M. arctoides*) macaques. *International Journal of Primatology* 7: 507–517.

**40** In at least one rat species: D. Q. Estep. 1988. Copulations by other males shorten the post-ejaculatory intervals of pairs of roof rats, *Rattus rattus*. *Animal Behaviour* 36: 299–300.

**40** The males of such species: A. P. Møller. 1988. Testis size, ejaculate quality, and sperm competition in birds. *Biological Journal of the Linnean Society* 33: 273–283.

**40** It has been found for mammals generally: A. P. Møller. Ejaculate quality, testis size and sperm production in mammals. *Functional Ecology* 3: 91–96; J. R. Ginsburg and D. I. Rubenstein. 1990. Sperm competition and variation in zebra mating behavior. *Behavioral Ecology and Sociobiology* 26: 427–434; R. L. Brownell and K. Ralls. 1986. Potential for sperm competition in baleen whales. *Reports of the International Whale Commission* (special Issue) 8: 97–112.

**40** Ditto for primates: A. P. Møller. 1988. Ejaculate quality, testis size and sperm competition in primates. *Journal of Human Evolution* 17: 479–483.

**41** The answer is pretty clear: T. Birkhead and A. P. Møller. 1992. *Sperm Competition in Birds*. San Diego: Academic Press.

**42** One researcher tried injecting: P. Sugawara. 1979. Stretch reception in the bursa copulatrix of the butterfly *Pieris rapae crucivora*, and its role in behaviour. *Journal of Comparative Physiology* 130: 191–199.

**42** It is interesting to note: R. E. Silberglied, J. G. Sheperd, and J. L. Dickinson. 1984. Eunuchs: the role of apyrene sperm in lepidoptera? *The American Naturalist* 123: 255–265; P. A. Cook and M. J. G. Gage. 1995. Effects of risks of sperm competition on the numbers of eupyrene and apyrene sperm ejaculated by the moth *Plodia interpunctella*. *Behavioral Ecology and Sociobiology* 36: 261–268.

**43** It is interesting that ring doves: C. J. Erickson and P. G. Zenone. 1976. Courtship differences in male ring doves: avoidance of cuckoldry? *Science* 192: 1353–1354.

**44** Copulating males use their penis: Jonathan K. Waage. 1979. Dual function of the damselfly penis: sperm removal and transfer. *Science* 203: 916–918; see also R. L. Smith, ed., 1984. *Sperm Competition and the Evolution of Animal Mating Systems.* New York: Academic Press.

**44** Male pygmy octopuses: J. A. Cigliano. 1995. Assessment of the mating history of female pygmy octopuses and a possible sperm competition mechanism. *Animal Behaviour* 49: 849–851.

**44** Prior to mating: P. L. Miller. 1990. Mechanisms of sperm removal and sperm transfer in *Orthetrum coerulescens* (Fabricus) (Odonata: Libellulidae). *Physiological Entomology* 15: 199–209.

**44** The more time a female dunnock: N. B. Davies. 1983. Polyandry, cloaca-pecking and sperm competition in dunnocks. *Nature* 302: 334–336.

**45** If and when the victim copulates: J. Carayon. 1974. Insemination traumatique heterosexualle et homosexualle chex *Xylocoris maculipennis. Comptes Rendues Academie de Sciences de Paris, Series D* 278: 2803–2806.

**45** To put it bluntly: Except if that someone is a close relative, in which case such behavior can be favored via "kin selection." For our purposes, however, it is accurate to say that any tendency to rear someone else's offspring will be strongly selected against.

**45** It may surprise some readers: F. Cezilly and R. G. Nager. 1995 Comparative evidence for a positive association between divorce and extra-pair paternity in birds. *Proceedings of the Royal Society of London, Series B* 262: 7–12.

**46** Among indigo buntings: D. F. Westneat and P. W. Sherman. 1990. When monogamy isn't. *The Living Bird Quarterly* 9: 24–28.

**47** So-called cooperative breeding: P. P. Rabenold, K. N. Rabenold, W. H. Piper, et al. 1990. Shared paternity revealed by genetic analysis in cooperative breeding tropical wrens. *Nature* 348: 538–542.

**47** It turned out that it didn't matter: O. Svensson, C. Magnhagen, E. Forsgren, and C. Kvarnemo. 1998. Parental behaviour in relation to the occurrence of sneaking in the comon goby. *Animal Behaviour* 56: 175–179.

**47** It is not invariant: L. A. Whittingham and J. T. Lifjeld. 1995. High paternal investment in unrelated young: extra-pair paternity and male parental care in house martins. *Behavioral Ecology and Sociobiology* 37: 103–108.

**48** Or maybe in certain cases: I. P. F. Owens. 1993. When kids just aren't worth it: cuckoldry and parental care. *Trends in Ecology and Evolution* 8: 269–271.

**48** The male behaved aggressively: D. P. Barash. 1976. The male response to apparent female adultery in the mountain bluebird, *Sialia currucoides:* An evolutionary interpretation. *The American Naturalist* 110: 1097–1101.

**48** In any event, there have since been: E.g., N. B. Davies, B. J. Hatchwell, T. Burke, and T. Robson. 1992. Paternity and parental effort in dunnocks *Prunella modularis*:

how good are male chick-feeding rules? *Animal Behaviour* 43: 729–745; L. A. Whittingham, P. D. Taylor, and R. J. Robertson. 1992. Confidence of paternity and male parental care. *The American Naturalist* 139: 1115–1125; B. C. Sheldon, K. Rasanen, and P. C. Dias. 1997. Certainty of paternity and parental care in collared flycatchers: an experiment. *Behavioral Ecology* 8: 421–428; see review by Jonathan Wright. 1998. Paternity and paternal care. In *Sperm Competition and Sexual Selection*, ed. T. R. Birkhead and A. P. Loller. San Diego: Academic Press.

**48** In another cooperatively breeding bird species: W. D. Koenig. 1990. Opportunity of parentage and nest destruction in polygynandrous acorn woodpeckers, *Melanerpes formicivorus*. *Behavioral Ecology* 1: 55–61.

**49** For example, male indigo buntings: D. F. Westneat. 1988. Parental care and extrapair copulations in the indigo bunting. *Auk* 105: 149–160.

**50** Their offspring get short shrift: D. P. Barash. 1975. Ecology of parental behavior in the hoary marmot (*Marmota caligata*). *Journal of Mammalogy* 56: 613–618.

**50** Attractive males usually provide: A. P. Møller and R. Thornhill. 1998. Male parental care, differential parental investment by females and sexual selection. *Animal Behaviour* 55: 1507–1515.

**50** This tendency is captured: B. C. Sheldon, J. Merila, A. Qvarnstrom, L. Gustafsson, and H. Ellegren. 1997. Paternal contribution to offspring condition is predicted by size of male secondary sexual characteristic. *Proceedings of the Royal Society of London, Series B* 264: 297–302.

**51** Incidentally, long-tailed male barn swallows: A. P. Møller, A. Barbosa, J. J. Cuervo, F. de Lope, S. Merino, and N. Saino. 1998. Sexual selection and tail streamers in the barn swallow. *Proceedings of the Royal Society of London, Series B.* 265: 409–414.

**51** They simply have nothing: R. H. Wagner, M. D. Schug, and E. S. Morton. 1996. Confidence of paternity, actual paternity and parental effort by purple martins. *Animal Behaviour* 52: 123–132.

**52** I and others have documented: F. McKinney, S. R. Derrickson, and P. Mineau. 1983. Forced copulation in waterfowl. *Behaviour* 86: 250–294.

**52** Since in nearly all cases: A. P. Møller. 1987. House sparrow *Passer domesticus* communal displays. *Animal Behaviour* 35: 203–210.

**53** Males are sometimes aggressive: D. F. Westneat. 1987. Extrapair copulations in a predominatly monogamous bird: observations of behaviour. *Animal Behaviour* 35: 865–876.

**54** There is in fact: R. Thornhill and C. Palmer. 2000. *The Natural History of Rape.* Cambridge, MA: MIT Press.

**54** This has subsequently been found: J. T. Burns, K. Cheng, and F. McKinney. 1980. Forced copulation in captive mallards: I. Fertilization of eggs. *Auk* 97:

875–879; F. McKinney and P. Stolen. 1982. Extra-pair bond courtship and forced copulation among captive green-winged teal (*Anas carolinensis*). *Animal Behaviour* 30: 461–474; F. McKinney, K. M. Cheng, and D. Bruggers. 1984. Sperm competition in apparently monogamous birds. In *Sperm Competition and the Evolution of Animal Mating Systems,* ed. R. L. Smith. New York: Academic Press; K. M. Cheng, J. T. Burns, and F. McKinney. 1983. Forced copulation in captive mallards: III. Sperm competition. *Auk* 100: 302–310.

54 Probably because, given "last male advantage": L. G. Sorenson. 1994. Forced extra-pair copulation and mate guarding in the white-cheeked pintail: timing and tradeoffs in an asynchronously breeding duck. *Animal Behaviour* 48: 519–533.

54 Research conducted at the largest known goose colony: P. O. Dunn, A. D. Afton, M. L. Gloutney, and R. T. Alisauskas. 1999. Forced copulation results in few extrapair fertilizations in Ross's and lesser snow geese. *Animal Behaviour* 57: 1071–1081.

54 It appears that species: G. Gauthier. 1988. Territorial behavior, forced copulations and mixed reproductive strategy in ducks. *Wildfowl* 39: 102–114.

55 About one-third of spousal killings: M. Daly, M. Wilson, and S. Weghorst. 1982. Male sexual jealousy. *Ethology and Sociobiology* 3: 11–27.

55 The frequency of infidelity-generated violence: J. M. Tanner. 1970. *Homicide in Uganda, 1964.* Uppsala, Sweden: Scandinavian Institute of African Studies; C. F. Lobban. 1972. *Law and Anthropology in the Sudan.* African Studies Seminar Series No. 13. Khartoum, Sudan: Sudan Research Unit, Khartoum University.

55 Furthermore, as we shall see: D. M. Buss and D. P. Schmitt. 1994. Sexual strategies theory: a contextual evolutionary analysis of human mating. *Psychological Review* 100: 204–232.

## CHAPTER 3 Undermining the Myth: Females (Choosing Male Genes)

57 Clearly, there was some hanky-panky: O. Bray, J. Kennelly, and J. Guarlno. 1975. Fertility of eggs produced on territories of vasectomized red-winged blackbirds. *Wilson Bulletin* 87: 187–195.

58 In such cases—and especially when genetic testing: S. M. Smith. 1988. Extra-pair copulations in black-capped chickadees: the role of the female. *Behaviour* 107: 15–23; B. Kempenaers, G. R. Verheyen, M. Van den Broeck, T. Burke, C. Van Broeckhoven, and A. A. Dhondt. 1992. Extra-pair paternity results from female preference for high-quality males in the blue tit. *Nature* 357: 494–496.

58 One possibility is that extra-pair copulations: T. Halliday and S. Arnold. 1987. Multiple mating by females: a perspective from quantitative genetics. *Animal Behaviour* 35: 939–941.

59 An interesting idea, this: K. M. Cheng and P. B. Siegel. 1990. Quantitative genetics of multiple mating. *Animal Behaviour* 40: 406–407.

**60** The upshot is that females who copulate: D. W. Pyle and M. H. Gromko. 1978. Repeated mating by female *Drosophila melanogaster:* the adaptive importance. *Experimentia* 34: 449–450; T. R. Birkhead and A. P. Møller. 1992. *Sperm Competition in Birds: Evolutionary Causes and Consequences.* London: Academic Press.

**60** This is suggested by the fact: J. Graves, J. Ortega-Ruano, and P. J. B. Slater. 1993. Extra-pair copulations and paternity in shags: do females choose better males? *Proceedings of the Royal Society of London, Series B* 253: 3–7.

**60** Females consistently have more EPCs: J. Wetton and D. Parkin. 1991. An association between fertility and cuckoldry in the house sparrow, *Passer domesticus. Proceedings of the Royal Society of London, Series B* 245: 227–233.

**60** Among those birds whose females: E. M. Gray. 1997. Do red-winged blackbirds benefit genetically from seeking copulations with extra-pair males? *Animal Behaviour* 53: 605–623.

**61** They do not initiate: D. F. Westneat. 1992. Do female red-winged blackbirds engage in a mixed mating strategy? *Ethology* 92: 7–28.

**61** Extra-pair copulations are frequent: J. H. Wetton and D. T. Parkin. 1991. An association between fertility and cuckoldry in the house sparrow *Passer domesticus. Proceedings of the Royal Society of London, Series B* 245: 227–233.

**61** In addition, litter size is larger: J. L. Hoogland. 1998. Why do Gunnison's prairie dogs copulate with more than one male? *Animal Behaviour* 55: 351–359.

**61** Not all mammals show this pattern: J. O. Murie. 1996. Mating behavior of Columbian ground squirrels: I. Multiple mating by females and multiple paternity. *Canadian Journal of Zoology* 73: 1819–1826.

**61** In fact, even in another prairie dog species: J. L. Hoogland. 1995. *The Black-Tailed Prairie Dog: Social Life of a Burrowing Mammal.* Chicago: University of Chicago Press.

**61** In some mammals, there is actually a *reduction*: K. E. Wynne-Edwards and R. D. Lisk. 1984. Djungarian hamsters fail to conceive in the presence of multiple males. *Animal Behaviour* 32: 626–628.

**62** The more males a female adder mates with: T. Madsen, R. Shine, J. Loman, and T. Hakansson. 1992. Why do female adders copulate so frequently? *Nature* 365: 440–441.

**63** DNA fingerprinting has shown: M. Ollson, R. Shine, A. Gullberg, A. Madsen, and J. Tegelstrom. 1996. Female lizards control the paternity of their offspring by selective use of sperm. *Nature* 383: 585.

**63** By multiple mating, a questing female: J. A. Zeh and D. W. Zeh. 1996. The evolution of polyandry: I. Intragenomic conflict and genetic incompatability. *Proceedings of the Royal Society of London, Series B* 263: 1711–1717; J. A. Zeh and D. W. Zeh. 1997. The evolution of polyandry: II. Post-copulatory defences against genetic incompatability. *Proceedings of the Royal Society of London, Series B* 264: 69–75.

**63** In a bird species wonderfully called: M. G. Brooker, I. Rowley, M. Adams, and P. Baverstock. 1990. Promiscuity: an inbreeding avoidance mechanism in a socially monogamous species? *Behavioral Ecology and Sociobiology* 26: 191–199.

**63** Interestingly, when female mice: W. K. Potts, C. J. Manning, and E. K. Wakeland. 1991. Mating patterns in semi-natural populations of mice influenced by MHC genotype. *Nature* 352: 619–621.

**63** Female primates, for their part: B. A. Smuts. 1987. Gender, aggression and influence. In *Primate Societies,* ed. B. Smuts, D. L. Cheney, R. M. Seyfarth, R. W. Wrangham, and T. T. Struhsaker. Chicago: University of Chicago Press.

**64** For example, in one troop: Y. Takahata. 1982. The socio-sexual behavior of Japanese monkeys. *Zeitschrift fur Tierpsychologie* 59: 89–108.

**64** In one remarkable case: R. Sekulic. 1982. Behavior and ranging patterns of a solitary female red howler (*Alouatta seniculus*). *Folia Primatologica* 38: 217–232.

**64** A key summary point is that: S. B. Hrdy and P. L. Whitten. 1987. Patterning of sexual activity. In *Primate Societies,* ed. B. Smuts, D. L. Cheney, R. M. Seyfarth, R. W. Wrangham, and T. T. Struhsaker. Chicago: University of Chicago Press.

**64** In this species, therefore: O. Ratti, M. Hovi, A. Lundberg, H. Tegelstrom, and R. Alatalo. 1995. Extra-pair paternity and male characteristics in the pied flycatcher. *Behavioral Ecology and Sociobiology* 37: 419–425.

**66** They also vocalize loudly: C. R. Cox and B. J. LeBoeuf. 1977. Female incitation of male competition: a mechanism in sexual selection. *The American Naturalist* 111: 317–335; J. H. Poole. 1989. Mate guarding, reproductibve success and female choice in African elephants. *Animal Behaviour* 37: 842–849.

**66** Or females can advertise: R. H. Wiley and J. Poston. 1996. Indirect mate choice, competition for mates, and co-evolution of the sexes. *Evolution* 50: 1371–1381.

**66** In eight of twelve observed copulations: J. J. Perry-Richardson, C. S. Wilson, and N. B. Ford. 1990. Courtship of the garter snake, *Thamnophic marianus*, with a description of a female behavior for coitus interruption. *Journal of Herpetology* 24: 76–78.

**66** Interestingly, of ten such forced copulations: R. Thornhill. 1988. The jungle fowl hen's cackle incites male competition. *Verhalten Deutsche Zoologische Geselschaft* 81: 145–154.

**66** Once, after he forced such a copulation: T. R. Birkhead and A. P. Møller. 1992. *Sperm Competition in Birds: Evolutionary Causes and Consequences.* London: Academic Press.

**67** For example, a female pied flycatcher: M. Hovi and O. Ratti. 1994. Mate sampling and assessment procedures in female pied flycatchers (*Ficedula hypoleuca*). *Ethology* 96: 127–137.

67 In this situation, most females: C. T. Gabor and T. R. Haliday. 1997. Sequential mate choice by smooth newts: females become more choosy. *Behavioral Ecology* 8: 162–166.

68 Not only that, but the offspring: P. J. Watson. 1998. Multi-male mating and female choice increase offspring growth in the spider *Neriene litigiosa* (Linyphiidae). *Animal Behaviour* 55: 387–403.

68 For example, among waterfowl: F. McKinney, S. R. Derrickson, and P. Mineau. 1983. Forced copulation in waterfowl. *Behaviour* 86: 250–294.

69 This seems to be a way for females: J. A. Zeh, S. D. Newcomer, and D. W. Zeh. 1998. Polyandrous females discriminate against previous mates. *Proceedings of the National Academy of Sciences* 95: 13732–13736.

69 It had already been demonstrated: J. A. Zeh. 1997. Polyandry and enhanced reproductive success in the harlequin beetle-riding pseudoscorpion. *Behavioral Ecology and Sociobiology* 40: 111–118.

69 In one type of beetle: M. S. Archer and M. E. Elgar. 1999. Female preference for multiple partners: sperm competition in the hide beetle, *Dermestes maculatus*. *Animal Behaviour* 58: 669–675.

69 In some insects, the female deposits: J. A. Zeh, S. D. Newcomer, and D. W. Zeh. 1998. Polyandrous females discriminate against previous mates. *Proceedings of the National Academy of Sciences* 95: 13732–13736.

69 Most commonly, however, it appears that: M. Petrie. 1994. Improved growth and survival of offspring of peacocks with more elaborate trains. *Nature* 341: 598–599.

70 Male house sparrows, for example: A. P. Møller. 1990. Sexual behaviour is related to badge size in the house sparrow *Passer domesticus*. *Behavioral Ecology and Sociobiology* 27: 23–29.

70 Among zebra finches: N. Burley and D. Price. 1991. Extra-pair copulation and attractiveness in zebra finches. *Proceedings of the International Ornithological Congress* 20: 1367–1372.

71 This supports the hypothesis: R. Wagner. 1991. The role of extra-pair copulations in razorbill mating strategies. D. Phil. thesis, University of Oxford, Oxford, UK.

71 As a result of such choice: P. Dunn and A. Cockburn. 1996. Evolution of male paternal care in a bird with almost complete cuckoldry. *Evolution* 50: 2542–2548.

72 Incidentally, these helpers: Ibid.

72 Accordingly, females are well advised: Allison Welch, R. Semlitsch, and H. C. Gerhardt. 1998. Call duration as an indicator of genetic quality in male gray tree frogs. *Science* 280: 1928–1930.

72 It has been suggested that bright coloration: W. D. Hamilton. 1990. Mate choice near or far. *American Zoologist* 30: 341–352.

73 Such a correlaion does not: A. P. Møller. 1997. Immune defence, extra-pair paternity, and sexual selection in birds. *Proceedings of the Royal Society of London, Series B* 264: 561–566.

73 It is then also revealing: B. Kempenaers, G. R. Verheyen, M. Van den Broeck, T. Burke, C. Van Broeckhoven, and A. A. Dhondt. 1992. Extra-pair paternity results from female preference for high-quality males in the blue tit. *Nature* 357: 494–496.

73 Of the five males that died: Ibid.

73 Among blue tits, for example: Ibid.

74 Dominant male cattle egrets: M. Fujioka and S. Yamagishi. 1981. Extramarital and pair copulations in the cattle egret. *Auk* 98: 134–144; P. C. Frederick. 1987. Extrapair copulations in the mating system of white ibis (*Eudocimus albus*). *Behaviour* 100: 170–201.

74 Not only that, but EPCs: S. M. Smith. 1988. Extra-pair copulations in black-capped chickadees: the role of the female. *Behaviour* 107: 15–23.

74 Accordingly, older male red-winged blackbirds: P. J. Weatherhead and P. T. Boag. 1995. Pair and extra-pair mating success relative to male quality in red-winged blackbirds. *Behavioral Ecology and Sociobiology* 37: 81–91.

74 A similar age-related pattern: E. Roskaft. 1983. Male promiscuity and female adultery by the rook *Corvus frugilegus Ornis Scandinavica* 14: 175–179.

75 In the previous chapter, we encountered: D. Hasselquist, S. Bensch, and T. von Schantz. 1996. Correlation between male song repertoire, extra-pair paternity and offspring survival in the great reed warbler. *Nature* 381: 229–232.

75 In the case of bilaterally symmetrical creatures: Leigh Van Valen. 1962. A study of fluctuating asymmetry. *Evolution* 16: 125–142; P. A. Parsons. 1990. Fluctuating asymmetry: an epigenetic measure of stress. *Biological Reviews* 65: 131–145.

76 One was titled: A. Møller. 1992. Female swallow preference for symmetrical male sexual ornament. *Nature* 357: 238–240

76 The second study was titled: A. Møller. 1992. Parasites differentially increase the degree of fluctuating asymmetry in secondary sexual characteristics. *Journal of Evolutionary Biology* 5: 691–700.

76 It does: More symmetry equals better looking: S. W. Gangestad, R. Thornhill, and R. A. Yeo. 1994. Facial attractiveness, developmental stability, and fluctuating asymmetry. *Ethology and Sociobiology* 15: 73–85.

76 Not only that, but symmetrical men: R. Thornhill and S. W. Gangestad. 1994. Fluctuating asymmetry and human sexual behavior. *Psychological Science* 5: 297–302.

76 Women even report more orgasms: R. Thornhill, S. W. Gangestad, and R. Comer. 1996. Human female orgasm and mate fluctuating asymmetry. *Animal Behaviour* 50: 1601–1615.

77 So, when it comes to already-paired females: S. W. Gangestad and R. Thornhill. 1997. The evolutionary psychology of extrapair sex: the role of fluctuating asymmetry. *Evolution and Human Behavior* 18: 69–88.

77 The suggestion has even been made: G. F. MIller. 2000. *The Mating Mind*. New York: Doubleday.

78 As a result, a female is well advised: P. J. Weatherhead and R. J. Robertson, 1979. Offspring quality and the polygyny threshold: "The sexy son hypothesis." *The American Naturalist* 113: 201–208.

78 But the offspring of preferred males: T. M. Jones, R. J. Quinnell, and A. Balmford. 1998. Fisherian flies: the benefits of female choice in a lekking sandfly. *Proceedings of the Royal Society of London, Series B* 265: 1–7.

79 When researchers from the University of Oslo: E. Cunningham and T. Birkhead. 1997. Female roles in perspective. *Trends in Ecology and Evolution* 12: 337–338.

79 Not only that, but younger females: L. A. Dugatkin and J. Godin. 1992. Reversal of female mate choice by copying in the guppy (*Poecilia reticulata*). *Proceedings of the Royal Society of London, Series B* 249: 179–184.

79 Males swallows whose forked tails: A. P. Møller. 1994. *Sexual Selection and the Barn Swallow*. New York: Oxford University Press.

79 At the same time, the number of biological young: N. Saino, C. P. Rimmer, H. Ellegren, and A. P. Møller. 1997. An experimental study of paternity and tail ornamentation in the barn swallow (*Hirundo rustica*) *Evolution* 51: 562–570.

79 Males engaging in EPCs: A. P. Møller. 1988. Badge size in the house sparrow *Passer domesticus*: effects of intra- and intersexual selection. *Behavioral Ecology and Sociobiology* 22: 373–378.

80 A comparable finding applies: N. T. Burley, D. A. Enstrom, and L. Chitwood. 1994. Extra-pair relations in zebra finches: differential male success results from female tactics. *Animal Behaviour* 48: 1031–1041.

80 Again, red is desirable: N. T. Burley, P. G. Parker, and K. Lundy. 1996. Sexual selection and extrapair fertilization in a socially monogamous passerine, the zebra finch (*Taeniopygia guttata*). *Behavioral Ecology* 7: 218–226.

80 The findings are that: E.g., R. G. Edwards. 1955. Selective fertilization following the use of sperm mixtures in the mouse. *Nature* 175: 215–223; R. A. Beatty. 1951. Fertility and mixed semen from different rabbits. *Journal of Reproduction and Fertility* 1: 52–60; P. A. Martin and P. J. Dziuk. 1977. Assessment of relative fertility of males (cockerels and boars) by competitive mating. *Journal of Reproduction and Fertility* 49: 323–329.

81 Columbian ground squirrel females: J. O. Murie. 1995. Mating behavior of Columbian ground squirrels: I. Multiple mating by females and multiple paternity. *Canadian Journal of Zoology* 73: 1819–1826.

81 In another species of ground squirrel: D. E. Boellstorff, D. H. Owings, M. C. T. Penedo, and M. J. Hersek. 1994. Reproductive behaviour and multiple paternity of California ground squirrels. *Animal Behaviour* 47: 1057–1064.

82 In one study, detailed measurements: A. F. Dixson and N. I. Mundy. 1994. Sexual behaviour, sexual swelling and penile evolution in chimpanzees (*Pan troglodytes*). *Archives of Sexual Behavior* 23: 267–280.

82 In an article entitled: T. Birkhead, A. Møller, and W. J. Sutherland. 1993. Why do females make it so difficult to fertilize their eggs? *Journal of Theoretical Biology* 161: 51–60.

83 It emphasizes, as has been traditional: W. G. Eberhard. 1998. Female roles in sperm competition. In *Sperm Competition and Sexual Selection,* ed. T. Birkhead and A. Møller. San Diego: Academic Press.

83 But in fact, as Eberhard points out [subsequent quote]: Ibid.

83 William Eberhard, one last time: W. Eberhard. 1994. Evidence for widespread courtship during copulation in 131 species of insects and spiders. *Evolution* 48: 711–733.

84 Tempting as it is to attribute this: L. W. Simmons, P. Stockley, R. L. Jackson, and G. A. Parker. 1996. Sperm competition or sperm selection: no evidence for female influence over paternity in yellow dung flies *Scatophagia stercoraria*. *Behavioral Ecology and Sociobiology* 38: 199–206.

85 According to master myth-recounter Thomas Bulfinch [subsequent quote]: T. Bulfinch. 1855/1934. *Bulfinch's Mythology*. New York: Modern Library.

## CHAPTER 4 Undermining the Myth: Females (Other Considerations)

89 In a now-classic and much-cited study: A. J. Bateman. 1948 Intra-sexual selection in *Drosophila*. *Heredity* 2: 349–368.

89 Indian crested porcupines copulate: Z. Sever and J. Mendelssohn. 1988. Copulation as a possible mechanism to maintain monogamy in porcupines, *Hystrix indica*. *Animal Behaviour* 36: 1541–1542.

90 Newlywed kittiwake gulls copulate: I. W. Chardine, cited in Hunter et al. 1993. Why do females copulate repeatedly with one male? *Trends in Ecology and Evolution* 8: 21–26.

91 Red-billed gulls, studied off the coast of New Zealand: J. A. Mills. 1994. Extra-pair copulations in the red-billed gull: females with high-quality, attentive males resist. *Behaviour* 128: 41–64.

92 The technical article describing this system: L. Wolf. 1975. Prostitution behavior in a tropical hummingbird. *Condor* 77: 140–144.

**93** Indeed, only hive-owning males: E. W. Cronin and P. W. Sherman. 1977. A resoure-based mating system: the orange-rumped honeyguide. *The Living Bird Quarterly* 15: 5–32.

**93** Although there is little obvious benefit: M. A. Elgar. 1992. Sexual cannibalism in spiders and other invertebrates. In *Cannibalism: Ecology and Evolution Among Diverse Taxa*, ed. M. E. Elgar and B. J. Crespi. Oxford: Oxford University Press.

**93** Male moths attract females: T. Eisner and J. Meinwald. 1995. The chemistry of sexual selection. *Proceedings of the National Academy of Sciences* 92: 50–55.

**94** The results were startling: P. Gagneux, D. S. Woodruff, and C. Boesch. 1997. Furtive mating in female chimpanzees. *Nature* 387: 358–359.

**95** An ethically troubling discovery: S. B. Hrdy. 1979. *The Langurs of Abu.* Cambridge, MA: Harvard University Press.

**96** This may be quite important: T. Nishida and K. Kawanaka. 1985. Within-group cannibalism by adult male chimpanzees. *Primates* 26: 274–284.

**96** She has since expanded that notion: S. B. Hrdy. 1981. *The Woman That Never Evolved.* Cambridge, MA: Harvard University Press.

**96** It is well documented that among many primates: B. B. Smuts. 1987. Sexual competition and mate choice. In *Primate Societies*, ed. D. Cheney, R. Seyfarth, B. Smuts, R. Wrangham, and T. Struhsaker. Chicago: University of Chicago Press.

**96** University of Michigan primatologist Barbara Smuts: B. B. Smuts. 1985. *Sex and Friendship in Baboons.* Hawthorne, NY: Aldine.

**96** Given that, among primates in particular: D. P. Barash. 2001. *Revolutionary Biology: The new, gene-centered view of life.* New Brunswick, NJ: Transaction.

**97** So, if you are a male barn swallow: J. J. Soler, J. S. Cuervo, A. P. Møller, and F. de Lope. 1998. Nest building is a sexually selected behaviour in the barn swallow. *Animal Behaviour* 56: 1435–1444.

**97** Female appreciation of males: E. Forsgren. 1997. Female sand gobies prefer good fathers over dominant males. *Proceedings of the Royal Society of London, Series B* 264: 1283–1286.

**97** Not coincidentally, male dunnocks: N. B. Davies, I. R. Hartley, B. J. Hatchwell, and N. E. Langmore. 1996. Female control of copulations to maximize male help: a comparison of polygynandrous alpine accentors, *Prunella collaris*, and dunnocks *P. modularis*. *Animal Behaviour* 51: 27–47.

**98** Probably she does this to convince: N. B. Davies. 1983. Polyandry, cloaca pecking and sperm competition in dunnocks. *Nature* 302: 334–336.

**98** His sperm pass with extraordinary speed: M. Kohda, M. Tanimura, M. Kikue-Nakamura, and S. Yamagishi. 1995. Sperm drinking by female catfishes: a novel mode of insemination. *Environmental Biology of Fish* 42: 1–6.

**98** Most likely, the males were duped: J. O. Gjershaug, T. Jarvi, and E. Roskaft. 1989. Marriage entrapment by "solitary" mothers: a study on male deception by female pied flycatchers. *The American Naturalist* 133: 273–276.

**99** But they might nonetheless coerce an EPC: D. F. Westneat. 1992. Do female red-winged blackbirds engage in a mixed mating strategy? *Ethology* 92: 7–28.

**99** The researcher reports [subsequent quote]: P. C. Frederick. 1987. Extrapair copulations in the mating system of white ibis (*Eudocimus albus*). *Behaviour* 100: 170–201.

**100** Interestingly, this behavior was not characteristic: Ibid.

**100** Only after females finally break down: P. J. Watson. 1993. Foraging advantage of polyandry for female Sierra dome spiders (*Linyphia litigiosa: linyphiidae*) and assessment of alternative direct benefit hypotheses. *The American Naturalist* 141: 440–465.

**100** With growing awareness of the importance of EPCs: R. E. Ashcroft. 1976. A function of the pairbond in the common eider. *Wildfowl* 27: 101–105.

**100** No one knows the mechanism involved: E. J. A. Cunningham. 1997. Forced copulation and sperm competition in the mallard *Anas platyrhunchos*. Ph.D. dissertation, University of Sheffield, Sheffield, UK.

**101** "Peahens lay more eggs,": E.g., M. Petrie and A. Williams. 1993. Peahens lay more eggs for peacocks with larger trains. *Proceedings of the Royal Society of London, Series B* 251: 127–131; N. Burley. 1988. The differential allocation hypothesis: an experimental test. *The American Naturalist* 132: 611–628.

**102** Thus, of 11 heterosexual pairs observed: R. Palombit. 1994. Dynamic pair bonds in hylobatids: implications regarding monogamous social systems. *Behaviour* 128: 65–101.

**102** There are many possible reasons for pairs: S. Choudhury. 1995. Divorce in birds: a review of the hypotheses. *Animal Behaviour* 50: 413–429.

**102** Thus, a renowned study of cliff-nesting gulls: J. C. Coulson. 1972. The significance of the pair-bond in the kittiwake. In *Proceedings of the International Ornithology Congress*. Leiden, Holland: Brill.

**102** An alternative view, recently advanced: B. Ens, U. N. Safriel, and M. P. Harris. 1993. Divorce in the long-lived and monogamous oystercatcher, *Haematopus ostralegus*: incompatibility or choosing the better option? *Animal Behaviour* 45: 1199–1217; A. A. Dhondt and F. Adriaensen. 1994. Causes and effects of divorce in the blue tit *Parus caeruleus. Journal of Animal Ecology* 63: 979–987; M. Orell, S. Rytkonen, and K. Koivula. 1994. Causes of divorce in the monogamous willow tit, *Parus montanus*, and consequences for reproductive success. *Animal Behaviour* 48: 1143–1150.

**102** A research paper titled: D. Heg, B. Ens, R. T. Burke, L. Jenkins, and J. P. Krujit. 1993. Why does the typically monogamous oystercatcher (*Haematopus ostralegus*) engage in extra-pair copulations? *Behaviour* 126: 247–288.

**102** Assuming that most, if not all: M. S. Sullivan. 1994. Mate choice as an information gathering process under time constraint: implications for behaviour and signal design. *Animal Behaviour* 47: 141–151.

**102** Similarly, in another bird species: M. A. Colwell and L. W. Oring. 1989. Extra-pair mating in the spotted sandpiper: a female mate acquisition tactic. *Animal Behaviour* 38: 675–684; R. H. Wagner. 1991. The use of extrapair copulations for mate appraisal by razorbills, *Alca torda. Behavioral Ecology* 2: 198–203.

**103** They may even be increasing their chances: H. L. Gibbs, P. J. Weatherhead, P. T. Boag, B. N. White, L. M. Tabak, and K. J. Hoysak. 1990. Realized reproductive success of polygynous red-winged blackbirds revealed by DNA markers. *Science* 250: 1394–1397.

**103** It has also been suggested: R. H. Wagner. 1993. The pursuit of extra-pair copulations by female birds: a new hypothesis of colony formation. *Journal of Theoretical Biology* 163: 333–346.

**103** When eastern bluebirds have already succeeded: P. A. Gowaty and W. C. Bridges. 1991. Behavioral, demographic, and environmental correlates of extra-pair fertilizations in eastern bluebirds, *Sialia sialis. Behavioral Ecology* 2: 339–350.

**103** In bird species as diverse: A. D. Afton. 1985. Forced copulation as a reproductive strategy of male lesser scaup: a field test of some predictions. *Behaviour* 92: 146–167; D. Westneat. 1987. Extra-pair copulations in a predominantly monogamous bird: observations of behaviour. *Animal Behaviour* 35: 865–876.

**103** Thus, if divorce is more likely: F. Cezily and R. G. Nager. 1995. Comparative evidence for a positive association between divorce and extra-pair paternity in birds. *Proceedings of the Royal Society of London, Series B* 262: 7–12.

**104** The result was that fewer than half: E. Peterson, T. Jarvi, J. Olsen, J. Mayer, and M. Hedenskog. 1999. Male–male competition and female choice in brown trout. *Animal Behaviour* 57: 777–783.

**105** It seems likely that EPCs: B. C. Sheldon. 1993. Sexually transmitted disease in birds: occurrence and evolutionary significance. *Philosophical Transactions of the Royal Society of London, Series B* 339: 491–497.

**105** That is, in this species: E. M. Gray. 1996. Female control of offspring paternity in a western population of red-winged blackbirds (*Agelius phoeniceus*). *Behavioral Ecology and Sociobiology* 38: 267–278.

**106** Once out of the male's field of vision: Recounted in R. R. Baker and M. A. Bellis. 1995. *Human Sperm Competition*. London: Chapman & Hall.

**106** One exception is my own research: D. P. Barash. 1976. Male response to apparent female adultery in the mountain bluebird (*Sialia currucoides*): an evolutionary interpretation. *The American Naturalist* 110: 1097–1101.

**107** Female barn swallows who "cain't": A. Møller. 1988. Paternity and paternal care in the swallow, *Hirunda rustica. Animal Behaviour* 36: 996–1005.

**107** When males reduce paternal care: S. Markman, Y. Yom-Tov, and J. Wright. 1995. Male parental care in the orange-tufted sunbird: behavioural adjustment in provisioning and nest guarding effort. *Animal Behaviour* 50: 655–669.

**108** In other cases, compensation: N. Saino and A. P. Møller. 1995. Testosterone-induced depression of male parental behavior in the barn swallow: female compensation and effects on seasonal fitness. *Behavioral Ecology and Sociobiology* 36: 151–157.

**108** Although eastern bluebirds are normally monogamous: P. A. Gowaty. 1983. Male parental care and apparent monogamy in eastern bluebirds (*Sialia sialis*). *The American Naturalist* 121: 149–157.

**108** The risk that males may reduce: R. A. Mulder, P. O. Dunn, A. Cockburn, K. A. Lazenby-Cohen, and M. J. Howell. 1994. Helpers liberate female fairy-wrens from constraints on extra-pair mate choice. *Proceedings of the Royal Society of London, Series B* 255: 223–229.

**108** Among these species, EPCs: J. W. Chardine. 1987. Influence of pair-status on the breeding behaviour of the kittiwake *Rissa tridactyla* before egg-laying. *Ibis* 129: 515–526; S. A. Hatch. 1987. Copulation and mate guarding in the northern fulmar. *Auk* 104: 450–461.

**108** Previous research has shown: A. P. Møller. 1991. Defence of offspring by male swallows, *Hirundo rustica*, in relation to participation in extra-pair copulations by their mates. *Animal Behaviour* 42: 261–267.

**108** Not only are resident males zealous: P. J. Weatherhead, R. Montgomerie, H. L. Gibbs, and P. T. Boag. 1994. The cost of extra-pair fertilizations to female red-winged blackbirds. *Proceedings of the Royal Society of London, Series B* 258: 315–320.

**109** A study conducted : D. F. Westneat. 1992. Do female red-winged blackbirds engage in a mixed mating strategy? *Ethology* 92: 7–28.

**109** The salient finding of one research effort: A. Dixon, D. Ross, S. L. C. O'Malley, and T. Burke. 1994. Paternal investment inversely related to degree of extra-pair paternity in the reed bunting. *Nature* 371: 698–700.

**109** After an EPC that took place: T. H. Birkhead and J. D. Biggins. 1987. Reproductive synchrony and extra-pair copulation in birds. *Ethology* 74: 320–334.

**109** On the other hand, older male purple martins: E. S. Morton, L. Forman, and M. Braun. 1990. Extrapair fertilizations and the evolution of colonial breeding in purple martins. *Auk* 107: 275–283.

**110** Neither of the extra-pair males helped: Recounted in T. Birkhead and A. P. Møller. 1992. *Sperm Competition in Birds*. San Diego: Academic Press.

**110** This, in turn, might result: D. P. Barash. 2001. *Revolutionary Biology: The New Gene-Centered View of Life*. New Brunswick, NJ: Transaction.

**110** And it has already been found: J. V. Briskie, C. T. Naugler, and S. M. Leech. 1994. Begging intensity of nestling birds varies with sibling relatedness. *Proceedings of the Royal Society of London, Series B* 258: 73–78.

**110** For example, male dunnocks shadow: N. B. Davies. 1985. Cooperation and conflict among dunnocks, *Prunella modularis*, in a variable mating system. *Animal Behaviour* 33: 628–648.

**111** There already exists a phrase: M. C. McKitrick. 1990. Genetic evidence for multiple parentage in eastern kingbirds (*Tyrannus tyrannus*). *Behavioral Ecology and Sociobiology* 26: 149–155.

**111** Since they did not have their own nests: M. I. Sandell and M. Diemer. 1999. Intraspecific brood parasitism: a strategy for floating females in the European starling. *Animal Behaviour* 57: 197–202.

**111** Among the waterbirds known as coots: B. E. Lyon. 1993. Conspecific brood parasitism as a flexible female reproductive tactic in American coots. *Animal Behaviour* 46: 911–928.

**112** It appears that younger, weaker females: J. M. Eadie and J. M. Fryxell. 1992. Density dependence, frequency dependence, and alternative nesting strategies in goldeneyes. *The American Naturalist* 140: 621–64.

**112** Although it is relatively uncommon: Y. Yom-Tov, G. M. Dunnett, and A. Andersson. 1974. Intraspecific nest parasitism in the starling *Sturnus vulgaris*. *Ibis* 116: 87–90; F. McKinney. 1985. Primary and secondary male reproductrive strategies of dabbling ducks. In *Avian Monogamy*, ed. P. A. Gowaty and D. W. Mock. Washington, DC: American Ornithologists Union.

**112** One study found that among cliff swallows: C. F. Brown. 1984. Laying eggs in a neighbor's nest: benefit and cost of colonial nesting in swallows. *Science* 224: 518–519.

## CHAPTER 5 Why Does Monogamy Occur At All?

**114** Among kittiwake gulls: J. C. Coulson. 1966. The influence of the pair-bond and age on the breeding biology of the kittiwake gull, *Rissa tridactyla. Journal of Animal Ecology* 35: 269–279.

**114** For example, when males perceive themselves: N. T. Burley 1977. Parental investment, mate choice, and mate quality. *Proceedings of the National Academy of Sciences* 74: 3476–3479.

**115** In this situation, males: M. Milinski and T. C. M. Bakker. 1992. Costs influence sequential mate choice in sticklebacks, *Gasterosteus aculeatus. Proceedings of the Royal Society of London, Series B* 250: 229–233.

**115** In an earlier publication: M. Milinski and T. C. M. Bakker. 1990. Female sticklebacks use male coloration in mate choice and hence avoid parasitized males. *Nature* 344: 330–333.

**115** They are indicated by the title: S. Lopez. 1999. Parasitized female guppies do not prefer showy males. *Animal Behaviour* 57: 1129–1134.

**116** Dorothy Parker put it this way [subsequent quote]: D. Parker. 1936. *The Collected Poetry of Dorothy Parker.* New York: Modern Library.

**117** Males cannot monopolize: D. W. Tinkle. 1967. Home range, density, dynamics, and structure of a Texas population of the lizard, *Uta stansburiana.* In *Lizard Ecology,* ed. W. W. Mijstead. Columbia: University of Missouri Press.

**117** These animals may also be too aggressive: S. J. Hannon. 1984. Factors limiting polygyny in the willow ptarmigan. *Animal Behaviour* 32: 153–161.

**117** Similarly, among eastern bluebirds: P. A. Gowaty. 1983. Male parental care and apparent monogamy among eastern bluebirds (*Sialia sialis*). *The American Naturalist* 121: 149–157.

**118** When this is true: E.g., J. P. Lightbody and P. J. Weatherhead. 1988. Female settling patterns and polygyny: tests of a neutral mate choice hypothesis. *The American Naturalist* 132: 20–33; I. R. Hartley, M. Shepherd, and D. B. A. Thompson. 1995. Habitat selection and polygyny in breeding corn buntings *Miliaria calandra. Ibis* 137: 508–514.

**118** This aggression, incidentally: E. Cunningham and T. Birkhead. 1997. Female roles in perspective. *Trends in Ecology and Evolution* 12: 337–338.

**118** In starlings, at least: M. I. Sandell and J. G. Smith. 1996. Already mated females constrain male mating success in the European starling. *Proceedings of the Royal Society of London, Series B* 263: 742–747; M. I. Sandell and J. G. Smith. 1997. Female aggression in the European starling during the breeding season. *Animal Behaviour* 53: 13–23.

**118** In another bird species: N. E. Langmore and N. B. Davies. 1997. Female dunnocks use vocalizations to compete for males. *Animal Behaviour* 53: 881–890.

**118** Because mated females were aggressive: J. P. Veiga. 1992. Why are house sparrows predominantly monogamous? A test of hypotheses. *Animal Behaviour* 43: 361–370.

**118** Given that female house sparrows: J. P. Veiga. 1990, Infanticide by male and female house sparrows. *Animal Behaviour* 39: 496–502.

**118** A growing number of studies: T. Slagsvold, T. Amundsen, S. Dale, and H. Lampe. 1992. Female-female aggression explains polyterritoriality in male pied flycatchers. *Animal Behaviour* 43: 397–407.

**119** The feistiness of resident females beavers: H. E. Hodgdon and J. S. Larsen. 1973. Some sexual differences in behaviour within a colony of marked beavers (*Castor canadensis*). *Animal Behaviour* 21: 147–152.

**119** Primatologist Barbara Smuts: B. B. Smuts. 1987. Gender, aggression and influence. In *Primate Societies,* ed. D. Cheney, R. Seyfarth, B. Smuts, R. Wrangham, and T. Struhsaker. Chicago: University of Chicago Press.

119 There are also many cases: L. H. Frame and G. W. Frame. 1976. Female African wild dogs emigrate. *Nature* 263: 227–229; L. D. Mech. 1970. *The Wolf.* Garden City, NJ: Natural History Press; P. D. Moehlman. 1979. Jackal helpers and pup survival. *Nature* 277: 382–383; R. F. Ewer. 1973. The behaviour of the meerkat, *Suricata suricatta. Zeitschrift fur Tierpsychologie* 20: 570–607.

119 Only if this alpha female: J. M. Packard, U. S. Seal, L. D. Mech, and E. D. Plotka. 1985. Causes of reproductive failure in two family groups of wolves (*Canis lupus*). *Zeitschrift fur Tierpsychologie* 69: 24–40.

120 Thus liberated from their jealous wives: A.-K. Eggert and S. I. Sakaluk. 1995. Female coerced monogamy in burying beetles. *Behavioural Ecology and Sociobiology* 37: 147–154.

120 This suggests that what constrains: M. I. Sandell and H. G. Smith. 1996. Already mated females constrain male mating success in the European starling. *Proceedings of the Royal Society of London, Series B* 263: 743–747.

121 Equally interesting: M. Eens and R. Pinxten. 1995. Inter-sexual conflicts over copulation in the European starling: evidence for the female mate guarding hypothesis. *Behavioral Ecology and Sociobiology* 36: 71–81; M. Eens and R. Pinxten. 1996. Female European starlings increase their copulation solicitation rate when faced with the risk of polygyny. *Animal Behaviour* 51: 1141–1147.

121 In such cases: M. Petrie. 1992. Copulation frequency in birds: why do females copulate more than once with the same male? *Animal Behaviour* 44: 790–792.

121 Among blue tits, for example: B. Kempenaers, G. R. Verheyen, M. Van den Broeck, T. Burke, C. Van Broeckhoven, and A. A. Dhondt. 1992. Extra-pair paternity results from female preference for high-quality males in the blue tit. *Nature* 357: 494–496.

122 By mating repeatedly: M. Petrie et al. 1992. Multiple mating in a lekking bird: why do peahens mate with more than one male and with the same male more than once? *Behavioral Ecology and Sociobiology* 31: 349–358.

122 It is notable that female mate-guarding: E. Creighton. 2000. Female mate guarding: no evidence in a socially monogamous species. *Animal Behaviour* 59: 201–207.

123 The suggestion has been made: M. F. Small. 1988. Female primate sexual behavior and conception: are there really sperm to spare? *Current Anthropology* 29: 81–100.

123 In such cases, females often: K. Summers. 1990. Parental care and the cost of polygyny in the green dart-poison frog. *Behavioral Ecology and Sociobiology* 27: 307–313.

124 The predominant theoretical explanation: G. H. Orians. 1969. On the evolution of mating systems in birds and mammals. *The American Naturalist* 103: 589–603.

125 Male rats, for example: L. Krames and L. A. Mastromatteo. 1973. Role of olfactory stimuli during copulation in male and female rats. *Journal of Comparative and Physiological Psychology* 85: 528–535.

125 This, in turn, could help: E.g., S. J. Hannon and G. Dobush. 1997. Pairing status of male willow ptarmigan: is polygyny costly to males? *Animal Behaviour* 53: 369–380.

126 Here, without the social and sexual distractions: D. P. Barash. 1975. Ecology of paternal behavior in the hoary marmot: an evolutionary interpretation. *Journal of Mammalogy* 56: 612–615.

127 Whether monogamous or polygynous: D. M. B. Parish and J. C. Coulson. 1998. Parental investment, reproductive success and polygyny in the lapwing, *Vanellus vanellus*. *Animal Behaviour* 56: 1161–1167.

127 For a large number of women: See B. Ehrenreich. 1984. *The Hearts of Men*. New York: Doubleday.

127 Manucode monogamy: B. Beehler. 1985. Adaptive significance of monogamy in the trumpet manucode *Manucodia keraudrenii* (Aves: Paradisaeidae). In *Avian Monogamy*, ed. P. A. Gowaty and D. W. Mock. Washington, DC: American Ornithologists Union.

128 Sometimes, as in the water strider case: N. B. Davies. 1984. Sperm competition and the evolution of animal mating strategies. In *Sperm Competition and the Evolution of Animal Mating Systems*, ed. R. L. Smith. San Diego: Academic Press; L. Rowe, G. Arnqvist, A. Sih, and J. J. Krupa 1994. Sexual conflict and the evolutionary ecology of mating patterns: water striders as a model system. *Trends in Ecology and Evolution* 9: 289–293.

129 At this point, the male scorpionfly: R. Thornhill and K. P. Sauer. 1991. The notal organ of the scorpionfly (*Panorpa vulgaris*): an adaptation to coerce mating duration. *Behavioral Ecology* 2: 156–164.

129 These latter chemicals: T. Chapman, L. F. Liddle, J. M. Kalb, M. F. Wolfner, and L. Partridge. 1995. Cost of mating in *Drosophila melanogaster* females is mediated by male accessory gland products. *Nature* 373: 241–244.

129 After 41 generations: W. R. Rice. 1996. Sexually antagonistic male adaptation triggered by experimental arrest of female evolution. *Nature* 381: 232–234.

130 Hence, it may be worth noting: J. L. Koprowski. 1992. Removal of copulatory plugs by female tree squirrels. *Journal of Mammalogy* 73: 572–576.

130 Not to be entirely outdone: P. Stockley. 1997. Sexual conflict resulting from adaptations to sperm competition. *Trends in Ecology and Evolution* 12: 154–159.

131 Biologist Patricia Gowaty: P. A. Gowaty. 1996. Battles of the sexes and origins of monogamy. In *Partnerships in Birds: The Study of Monogamy*, ed. J. M. Black. Oxford: Oxford University Press.

**132** Recall the scene: L. van Valen. 1973. A new evolutionary law. *Evolutionary Theory* 1: 1–30.

**133** Nonetheless, even though paternal care: P. L. Whitten. 1987. Infants and adult males. In *Primate Societies*, ed. D. Cheney, R. Seyfarth, B. Smuts, R. Wranhgam, and T. Struhsaker. Chicago: University of Chicago Press.

**133** And even in these cases: M. M. West and M. J. Konner. 1976. The role of the father: an anthropological perspective. In *The Role of the Father in Child Development*, ed. M. E. Lamb. New York: Plenum Press.

**133** On the other hand: K. Hill and H. Kaplan. 1988. Tradeoffs in male and female reproductive strategies among the Ache: part 2. In *Human Reproductive Behaviour: A Darwinian Perspective*, ed. L. Betzig, M. Borgerhoff Mulder, and P. Turke. Cambridge: Cambridge University Press.

**134** There can also be fear: M. Daly and M. Wilson. 1988. *Homicide*. Hawthorne, NY: Aldine de Gruyter; M. Wilson and M. Daly. 1992. The man who mistook his wife for a chattel. In *The Adapted Mind: Evolutionary Psychology and the Generation of Culture*, ed. J. Barkow, L. Cosmides, and J. Tooby. New York: Oxford University Press.

**135** According to Hawkes: Reported in Michael Hagmann. 1999. More questions about the provider's role. *Science* 283: 777.

**137** Western Europe, however: R. D. Alexander. 1979. *Darwinism and Human Affairs*. Seattle: University of Washington Press.

**137** According to this view: Alexander especially emphasizes the role of external threats; for a focus on internal threats, see L. Betzig. 1986. *Despotism and Differential Reproduction*. Hawthorne, NY: Aldine de Gruyter.

**137** In any event, among Europeans in particular: K. MacDonald. 1995. The establishment and maintenance of socially imposed monogamy in Western Europe. *Politics and the Life Sciences* 14: 3–23.

## CHAPTER 6 What Are Human Beings, "Naturally"?

**141** And the most direct route: D. P. Barash and J. E. Lipton. 1997. *Making Sense of Sex*. Washington, DC: Island Press.

**141** Primates, too: R. D. Alexander, J. L. Hoogland, R. D. Howard, K. M. Noonan, and P. W. Sherman. 1979. Sexual dimorphism and breeding systems in pinnipeds, ungulates, primates, and humans. In *Evolutionary Biology and Human Social Behavior: An Anthropological Perspective*, ed. N. A. Chagnon and W. Irons. North Scituate, MA: Duxbury Press.

**141** After four years, both females: S. Biquand, A. Boug, V. Biquand-Guyot, and J. P. Gautier. 1994. Management of commensal baboons in Saudi Arabia. *Revue d'Ecology et de Biologie* 49: 213–222.

142 For example, in the primitive, egg-laying: P. D. Rismiller. 1992. Field observations on Kangaroo Island echidnas (*Tachyglossus aculeatus multiaculeatus*) during the breeding season. In *Platypus & Echidnas*, ed. M. L. Augee. Macquarie Centre, New South Wales, Australia: Royal Society of New South Wales.

143 And as time went on: D. M. Buss and D. P. Schmitt. 1993. Sexual strategies theory: an evolutionary perspective on human mating. *Psychological Review* 100: 204–232.

144 Men said that after just one day: Ibid.

144 Interestingly, among the 25 percent: R. D. Clark and E. Hatfield. 1989. Gender differences in receptivity to sexual offers. *Journal of Psychology and Human Sexuality* 2: 39–55.

145 Given this connection: D. Bar-Tal and L. Saxe. 1976. Perceptions of similarly attractive couples and individuals. *Journal of Personality and Social Psychology* 33: 772–782.

145 Willard Espy neatly expressed [subsequent quote]: W. R. Espy. 1998. *Skulduggery on Shoalwater Bay.* Windsor, Canada: Cranberry Press.

145 Although such standards: D. Buss. 1994. *The Evolution of Desire.* New York: Basic Books.

146 Some others appear to be monogamous: A. Fuentes. 1999. Re-evaluating primate monogamy. *American Anthropologist* 100: 890–907.

146 This view is no longer widely held [subsequent quote]: B. Malinowski. 1927. *Sex and Repression in Savage Society.* New York: Harcourt, Brace.

147 Of 185 human societies: C. S. Ford and F. Beach.1951. *Patterns of Sexual Behavior.* New York: Harper & Row.

147 The renowned anthropologist: G. P. Murdoch. 1949. *Social Structures.* London: Macmillan.

147 Anthropologist Weston LaBarre [subsequent quote]: W. LaBarre. 1954. *The Human Animal.* Chicago: University of Chicago Press.

148 "In Inca Peru, as probably everywhere" [subsequent quote]: L. L. Betzig. 1986. *Despotism and Differential Reproduction.* New York: Aldine.

149 Here is Twain's Devil [subsequent quote]: M. Twain. 1962. *Letters from the Earth.* New York: Harper & Row.

150 According to anthropologist Ruth Benedict [subsequent quote]: R. Benedict. 1934. *Patterns of Culture.* Boston: Houghton Mifflin.

151 As long as the situation is not unpleasant enough [subsequent quote]: Ibid.

151 And for good reason [subsequent quote]: Ibid.

152 After reviewing 116 different human societies: G. Broude. 1980. Extramarital sex norms in cross-cultural perspective. *Behavioral Science Research* 15: 181–218.

**152** Similarly, Laura Betzig evaluated: L. Betzig. 1989. Causes of conjugal dissolution: a cross-cultural study. *Current Anthropology* 30: 654–676.

**152** Friedrich Engels, in *The Origin of the Family*: F. Engels. 1942. *The Origin of the Family, Private Property and the State*. New York: International Publishers.

**152** In a famous oration: Demosthenes. 1992. *Apollodoros Against Neaira*. Westminster, UK: Aris & Phillips.

**154** Yet his wife reports [subsequent quote]: I. O. Reich. 1970. *Wilhelm Reich: A Personal Biography*. New York: Avon.

**154** In one of his many hundreds of letters: S. Freud, quoted in E. Jones. 1953. *The Life and Work of Sigmund Freud*, Vol 1. New York: Basic Books.

**155** His interpretation may be more appropriate [subsequent quote]: S. Freud. 1932. *The New Introductory Lectures*. London: Hogarth.

**156** In her description of the Dionysian inhabitants [subsequent quote]: R. Benedict. 1934. *Patterns of Culture*. Boston: Houghton Mifflin.

**158** Perhaps 40 percent of mammalian genera: D. Kleiman and J. Malcom. 1981. The evolution of male parental investment in mammals. In *Parental Care in Mammals*, ed. D. J. Gubernick and P. H. Klopfer. New York: Plenum Press.

**158** A review of 56 different human societies: G. E. Broude and S. J. Greene. 1976. Cross-cultural codes on twenty sexual attitudes and practices. *Ethnology* 15: 410–429.

**158** According to Kinsey and colleagues: A. C. Kinsey, W. B. Pomeroy, C. E. Martin, and P. H. Bebhard. 1953. *Sexual Behavior in the Human Female*. Philadelphia: W. B. Saunders.

**159** A different survey found that: L. Wolfe. 1981. *The Cosmo Report*. New York: Arbor House; Diagram Group. 1981. *Sex: A User's Manual*. London: Coronet Books.

**159** Interestingly, this was pretty much the optimum: K. Hill and H. Kaplan. 1988. Tradeoffs in male and female reproductive strategies among the Aché , Part 2. In *Human Reproductive Behavior*, ed. L. Betzig, M. Borgerhoff Mulder, and P. Turke, eds. Cambridge: Cambridge University Press.

**162** They estimate that in Great Britain: R. R. Baker and M. A. Bellis. 1995. *Human Sperm Competition*. London: Chapman & Hall.

**162** This is consistent with standard estimates: S. Macintyre and A. Sooman. 1992. Nonpaternity and prenatal genetic screening. *Lancet* 338: 839.

**162** If these data are reliable: R. R. Baker and M. A. Bellis. 1995. *Human Sperm Competition*. London: Chapman & Hall.

**163** This suggests that females: S. M. O'Connell and G. Cowlinshaw. 1994. Infanticide avoidance, sperm competition and female mate choice: the function of copulation calls in female baboons. *Animal Behaviour* 48: 687–694.

**163** The internal reproductive tract: M. Kanada, T. Daitoh, K. Mori, N. Maeda, K. Hirano, M. Irahara, T. Aono, and T. Mori. 1992. Etiological implication of autoantibodies to zona pellucida in human female infertility. *American Journal of Reproductive Immunology* 28: 104–109; K. Ahmad and R. K. Naz. 1992. Effects of human antisperm antibodies on development of preimplantation embryos. *Archives of Andrology* 29: 9–20.

**165** The only exception would be: G. A. Parker 1990. Sperm competition: sneaks and extra-pair copulations. *Proceedings of the Royal Society of London, Series B* 242: 127–133.

**167** At the same time, chemicals: C. Lindholmer. 1973. Survival of human sperm in different fractions of split ejaculates. *Fertility and Sterility* 24: 521–526.

**167** As Baker and Bellis put it: R. R. Baker and M. A. Bellis. 1995. *Human Sperm Competition*. London: Chapman & Hall.

**169** Yet most conceptions occur: C. E. Tutin. 1979. Mating patterns and reproductive strategies in a community of wild chimpanzees. *Behavioral Ecology and Sociobiology* 6: 29–38.

**169** Afterward, their sperm output: M. Freund. 1963. Effect of frequency of emission on semen output and an estimate of daily sperm production in man. *Journal of Reproduction and Fertility* 6: 269–286.

**169** By contrast, male chimps: J. Marson, D. Gervais, S. Meuris, R. W. Cooper, and P. Jouannet. 1989. Influence of ejaculation frequency on semen characteristics in chimpanzees. *Journal of Reproduction and Fertility* 85: 43–50.

**170** Harcourt concludes that sperm competition: A. H. Harcourt. 1991. Sperm competition and the evolution of nonfertilizing sperm in mammals. *Evolution* 45: 314–328.

**170** The male penchant for producing: G. A. Parker. 1982. Why are there so many tiny sperm? Sperm competition and the maintenance of two sexes. *Journal of Theoretical Biology* 96: 281–294.

**171** A similar pattern is found: J. T. Hogg. 1988. Copulatory tactics in relation to sperm competition in Rocky Mountain bighorn sheep. *Behavioral Ecology and Sociobiology* 22: 49–59.

**171** And it is interesting to note: A. H. Harcourt, P. H. Harvey, S. G. Larsen, and R. V. Short. 1981. Testis weight, body weight and breeding systems in primates. *Nature* 293: 55–57.

**171** Researchers at the University of Chicago: G. J. Wyckoff, W. Wang, and C.-I. Wu. 2000. Rapid evolution of male reproductive genes in the descent of man. *Nature* 403: 304–309.

**172** So wrote a prominent nineteenth-century physician: W. Acton. 1865. *Functions and Disorders of the Reproductive System*, 4th ed. London: Adams, Gold & Burt.

**172** "And we are waiting to be known": N. Angier. 1999. *Woman.* New York: Houghton Mifflin.

**173** Thus, evidence is now accumulating: M. A. Bellis and R. R. Baker. 1990. Do females promote sperm competition? Data for humans. *Animal Behaviour* 40: 997–999.

**173** They point to the suggestive finding: N. M. Morris and J. R. Udry. 1970. Variations in pedometer activity during the menstrual cycle. *Obstetrics and Gynecology* 35: 199–201.

**173** They expose more skin: K. Grammer, J. Dittami, and B. Fischmann. 1993. Changes in female sexual advertisement according to menstrual cycle. Paper presented at the International Congress of Ethology, Torremolinos, Spain.

**174** By spreading the breeding: A. G. Wilson and J. Terborgh. 1998. Cooperative polyandry and helping behavior in saddle-backed tamarins (*Saguinus fuscicollis*). In *Proceedings of the IXth Congress of the International Primatological Society.* Cambridge: Cambridge University Press.

**175** They are inclined to be socially monogamous: B. Kempenaers, G. R. Verheyen, M. Van den Broeck, T. Burke, C. Van Broeckhoven, and A. A. Dhondt. 1992. Extrapair paternity results from female preference for high-quality males in the blue tit. *Nature* 357: 494–496.

**175** This gives females the opportunity: S. B. Hrdy. 1986. Empathy, polyandry and the myth of the coy female. In *Feminist Approaches to Science*, ed. R. Bleier. New York: Pergamon.

**175** Other primates have a similar capacity: Reviewed in S. B. Hrdy and P. Whitten. 1986. The patterning of sexual activity. In *Primate Societies,* ed. D. Cheney, R. Seyfarth, B. Smuts, R. Wrangham, and T. Struhsaker. Chicago: University of Chicago Press.

**176** The reason for this effect is unknown: E. Kesseru. 1984. Sexual intercourse enhances the success of artificial insemination. *International Journal of Fertility* 29: 143–145.

**176** In short, social considerations: A. Troisi and M. Carosi. 1998. Female orgasm rate increases with male dominance in Japanese macaques. *Animal Behaviour* 56: 1261–1266.

**177** When a male bighorn sheep: V. Geist. 1971. *Mountain Sheep.* Chicago: University of Chicago Press.

**177** In one case, a female rhesus monkey: B. B. Smuts and R. W. Smuts. 1993. Male aggression and sexual coercion of females in nonhuman primates and other mammals: evidence and theoretical implications. *Advances in the Study of Behavior* 22: 1–63.

**178** If the errant female: H. Kummer. 1968. *Social Organization of Hamadryas Baboons.* Chicago: University of Chicago Press.

**178** When a rival male was introduced: D. Zumpe and R. P. Michael. 1990. Effects of the presence of a second male on pair-tests of captive cynomolgus monkeys (*Macaca fascicularis*): role of dominance. *American Journal of Primatology* 22: 145–158.

**178** It is prominent among chimps as well: J. Goodall. 1986. *The Chimpanzees of Gombe: Patterns of Behavior*. Cambridge, MA: Harvard University Press.

## CHAPTER 7 So What?

**182** According to Bertrand Russell [subsequent quote]: B. Russell. 1970. *Marriage and Morals*. New York: W. W. Norton.

**182** Hillary, Bishop of Poitiers. 1899. *St. Hillary of Poitiers, John of Damascus*. New York: Scribner's.

**184** Thus psychologist Havelock Ellis [subsequent quote]: H. Ellis. 1977. *Sex and Marriage: Eros in Contemporary Life*. Westport, CT: Greenwood Press.

**184** In their book *The Wandering Husband*: H. Spotnitz and L. Freeman. 1964. *The Wandering Husband*. Englewood Cliffs, NJ: Prentice-Hall.

**185** As an anonymous Greek author [subsequent quote]: From L. Untermeyer. 1956. *A Treasury of Ribaldry*. New York: Hanover House.

**185** According to Denis de Rougement, there is: D. de Rougemont. 1956. *Love in the Western World*. New York: Pantheon.

**185** According to de Rougement, we even have [subsequent quote]: Ibid.

**187** Not many people reflect on a spouse's adultery [subsequent quote]: J. Joyce. 1961. *Ulysses*. New York: Modern Library.

**190** It has been suggested that the mental health profession: F. Riesmann, J. Cohen, and A. Pearl. 1964. *Mental Health of the Poor*. New York: Free Press.

# Index

*Note*: Because certain words and concepts, such as adultery, cuckoldry, extra-pair copulations (EPCs), fidelity, marriage, monogamy, pair-bond, polygyny, and sex appear throughout, they are not specifically indexed here; to find them, read the book!